CAMBRIDGE

GUIDE TO
MINERALS
ROCKS AND FOSSILS

DISCARD

● A. C. BISHOP ● A. R. WOOLLEY ● W. R. HAMILTON ●

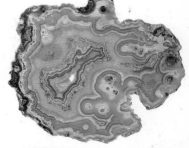

Preface to Second Edition

It is over 20 years since this book first appeared and numerous reprints and several foreign language editions are a measure of its success. The colour plates that were specially commissioned for the First Edition have been retained, although presented in a slightly different way, and are supplemented by 17 new plates. There have been significant advances in the geological sciences over the past two decades and the text has been thoroughly revised so as to take account of them. In particular we have tried to ensure that mineral names, chemical formulae and structural data accord to current international practice as recommended by the International Mineralogical Association (IMA). We are pleased to acknowledge the guidance given by such recent references as Hey's Mineral Index (Clark, 1993) and the Encyclopedia of Mineral Names by W. H. Blackburn and W. H. Dennen, edited by R. F. Martin (Canadian Mineralogist Special Publication 1, 1997). We are grateful to Dr Monica Grady for her help in revising and improving the section on meteorites and tektites. Some of the line drawings have been replaced, additional ones added, and colour has been introduced to improve clarity. The publishers have given the book a new look that we hope will appeal to readers.

Sadly, Dr Roger Hamilton, who was responsible for the part of the First Edition that dealt with fossils, died in 1979. Dr Brian Rosen, a colleague at The Natural History Museum, has undertaken the task of revising this section for the new edition. He has secured the help of several specialist contributors in order to ensure that the descriptions of fossils are accurate and up-to-date. These contributors are listed on page 223, and their help is most gratefully acknowledged.

We are pleased to acknowledge the links with The Natural History Museum in London in the preparation of this new edition. We are grateful also to Caroline Rayner, Stephen Scanlan and Laura Hill of Philip's for their help in the preparation and production of this volume.

A. C. Bishop, A. R. Woolley.

The additional colour photographs in this Edition were taken by The Photographic Unit of the Department of Exhibitions and Education, The Natural History Museum. The new line drawings have been prepared by Oxford Illustrators. Imperial measurements have been used throughout and a conversion table is given on the inside back cover.

PUBLISHED BY THE PRESS SYNDICATE OF THE UNIVERSITY OF CAMBRIDGE
The Pitt Building, Trumpington Street, Cambridge CB2 1RP, United Kingdom

CAMBRIDGE UNIVERSITY PRESS
The Edinburgh Building, Shaftesbury Road, Cambridge, CB2 2RU, United Kingdom
http://www.cup.cam.ac.uk
40 West 20th Street, New York, NY 10011–4211, USA
http://www.cup.org
10 Stamford Road, Oakleigh, Melbourne 3166, Australia

Previously published in Great Britain as the *Hamlyn Guide: Minerals, Rocks and Fossils*

This revised and expanded edition first published in 1999
by George Philip Limited

First published by Cambridge University Press in 1999

Printed in China

This edition only for sale in the United States of America and Canada

ISBN 0 521 77881 6 paperback

552

Contents

Rocks

Fossils

Introduction

This field guide is divided into three sections, namely minerals, rocks (including meteorites and tektites), and fossils. Each section comprises an introductory part, which is illustrated by line drawings, and a descriptive part, which is illustrated by line drawings and color photographs. The introductory sections include the minimum basic information required to follow the descriptive sections adequately, while the descriptive sections, for ease of reference, are always arranged so that photographs and accompanying text are closely adjacent.

To make the best use of the book the contents page and index should be used freely. The contents list will enable you to turn quickly to the appropriate section of the book, whereas if a tentative identification has been made, then reference to the index will immediately direct you to the relevant page. The index includes not only the names of specific minerals, rocks, and fossils, but also technical terms which are used in describing them. By consulting the index you will be referred to the page on which the term is defined, and possibly illustrated.

The stratigraphical column is given on page 328, and will be a particularly valuable reference for collectors of fossils.

How to collect in safety

Before setting out to collect it is most important to give thought to, and to take such precautions as would ensure, one's personal safety and preserve the interests of others. Excellent advice can often be obtained from local and national geologists' associations and societies. All those who contemplate geological fieldwork are urged to contact them and to follow their advice.

The basic equipment required to collect is a hammer, chisel, notebook and pencil, felt-tipped pen, wrapping materials, and a bag. The usual geological hammer has a square head and a chisel edge, which is particularly useful for splitting rocks when looking for fossils. Do not be tempted to use any other kind of hammer. Geological hammers are specifically tempered and others are likely to splinter when hammering, and metal splinters could damage the eyes. A steel chisel is sometimes required to pry open rocks which resist hammering, or for carefully breaking specimens which might be damaged by blows from a hammer. When hammering be very careful indeed of flying splinters of rock. Protective goggles can be obtained and should always be worn. Specimens should be carefully numbered; use either a felt-tipped pen or sticky tape on which a number can be written. The exact locality from which the specimens were collected should be recorded in the notebook. Specimens should always be wrapped in plenty of newspaper in order to prevent chipping or scratching, and small or delicate specimens are best carried in a small box, such as a match or cigar box. If a large collection is to be made, or if long distances are to be walked, then a stout backpack is the most suitable kind of bag to have.

The best places to collect minerals, rocks, and fossils are usually quarries, cliffs, road cuttings, and mine dumps, but any outcrop of rock may prove fruitful. It should be borne in mind that rock outcrops are potentially hazardous and appropriate protective clothing should be worn. In addition to the goggles mentioned above, a helmet of approved design gives protection against head injuries. It is always advisable to wear a helmet in quarries, and indeed mandatory in some places. Injuries can also result from rock falling on the feet. Boots, rather than trainers, or other soft footwear should be worn in the field, and those with protective toecaps offer the best protection. Particular care is needed, however, when

collecting near quarry faces or from the foot of cliffs, and permission must always be sought if it is intended to collect from outcrops on private land. An increasing number of sites are being designated as sites of special scientific interest, and given special protected status. Collecting from these sites may be restricted or forbidden: it is necessary to check in advance. Remember to take care and precautions if you intend to do field work on your own, and always tell someone of your intended route before setting out.

Geological maps, sometimes on a large scale, are available for most parts of the world, and they show the distribution and geological ages of the different rock types. This information should indicate where fossils are likely to be found, and where it is probably best to look for minerals, or for interesting rock types. If there is a museum in your area, a visit may well be worthwhile. Many museums not only have exhibits illustrating the geology of their vicinity, but they also usually have displays of minerals, fossils, and sometimes rocks, which will help you to "get your eye in", and give you some idea of what is to be found in your neighborhood.

Housing a collection

A collection is best kept in a cabinet of shallow drawers, with the specimens placed in individual cardboard trays. Under no circumstances should specimens be placed one on top of the other. Each specimen should have its own label giving details of what it is and where and when it was collected. A number should also be firmly glued to each specimen, and a corresponding entry made in a notebook, card index, or in a computer giving details such as name and locality, and any other relevant information. This entry is a safety precaution against accidental loss of, or damage to, the label attached to the specimen.

The system followed in this book will prove a useful guide in arranging specimens, though there are, of course, other systems which you may prefer to follow.

Mineral specimens, in particular, look their best when they are clean. To remove loose dust and dirt first take off the label, then immerse the specimen in clean water to which a little detergent has been added, and lightly scrub it with a soft brush. This should not be done, of course, with specimens which are soluble in water, or with very delicate material.

Further reading

Although in the introductory sections of this guide, outlines of the subjects of mineralogy, petrology (the study of rocks), and paleontology (the study of fossils) are given, it is obviously not possible in a single volume to do justice to these subjects. Although something like 600 specific types of mineral, rock, and fossil are described in the following pages, there are many other types which, for reasons of space, cannot be included. To help readers who would like to widen their knowledge, a list of recommended books is given on page 329. It would also be useful to include a list of the available geological maps and guides of particular areas, but such a list, if it is to be comprehensive, would need to be very long indeed. To find such maps and guides we suggest that you enquire at your local library.

For the real enthusiast there is no substitute for joining a geological society. Most countries have such societies organized on a national basis, but there are also local societies which cater mainly for the enthusiastic amateur, and which often have geological libraries, and organize field excursions to good collecting localities.

Minerals

The rocks which form the Earth, the Moon, and the planets are made up of minerals. Minerals are solid substances composed of atoms having an orderly and regular arrangement. This orderly atomic arrangement is the criterion of the crystalline state and it means also that it is possible to express the composition of a mineral as a chemical formula.

Crystals

When minerals are free to grow without constraint, they are bounded by crystal faces which are invariably disposed in a regular way such that there is a particular relationship between them in any one mineral species. A crystal is bounded by naturally formed plane faces, and its regular outward shape is an expression of its regular atomic arrangement.

The structure of minerals

The internal structure of minerals has been determined only during the 20th century, by the use of X-rays, although for about 200 years it had been appreciated that crystals are almost incredibly regular. This is not at once apparent for crystals of the same substance, such as quartz, which have faces that seem almost infinitely variable in their size and shape; it is only when the angles between corresponding pairs of faces are measured that the regularity becomes apparent. The angle between the same two faces in all crystals of the same mineral species is constant (Fig. 1).

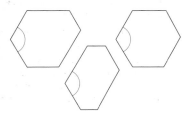

▲ Fig. 1 Constancy of interfacial angles

It is now known that this is because the constituent atoms pack together in a definite and orderly way. Crystals were studied long before this was appreciated, however, and from a study of external shape alone it was deduced that crystals were symmetrical and could be grouped according to their symmetry.

Crystal symmetry

We are familiar with symmetrical objects such as boxes, furniture, and even ourselves. Close inspection of such objects will reveal that they can be symmetrical about a *plane* such that if the object were to be cut in half along the plane, one half would be the mirror image of the other (Fig. 2). The human body is symmetrical externally about a vertical plane arranged from front to back.

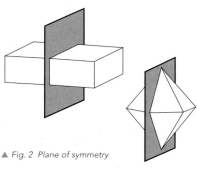

▲ Fig. 2 Plane of symmetry

Objects can also be symmetrical about a *line* or *axis* which is considered to pass through their center. When crystals are rotated about this axis they present the same appearance twice, three times, four times, or six times during a complete revolution (Fig. 3). The axis is called an axis of two-fold, three-fold, four-fold, or six-fold symmetry. Crystals never have an axis of five-fold symmetry. Finally, crystals can be said to be symmetrical about a *center* if a face on one side of the crystal has a corresponding parallel face on the other (Fig. 4).

The crystal systems

On the basis of their symmetry, crystals can be grouped into six crystal systems, and can be referred to imaginary reference axes, as shown in the diagrams (Fig. 5). A seventh crystal system, the trigonal, is recognized by many mineralogists. It has the same set of reference axes as the hexagonal system, but has a vertical three-fold axis of symmetry. These reference axes are chosen so as to be parallel to the edges of the unit cell (the repeat unit of pattern in a crystal structure), and hence they can be regarded as having length. The following table summarizes the seven crystal systems.

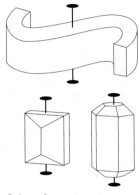

▲ Fig. 3 Axes of symmetry

▼ Fig. 4

Center of symmetry: cube and octahedron

No center of symmetry: tetrahedron

System	Symmetry	Reference axes
Cubic	Four three-fold axes	Three axes mutually at right-angles, and of equal length
Tetragonal	One vertical four-fold axis	Three axes mutually at right-angles; one axis conventionally held vertically, differing in length from the other two
Orthorhombic	Either one two-fold axis at the intersection of two mutually perpendicular planes; or three mutually perpendicular two-fold axes	Three axes mutually at right-angles, all of different length
Monoclinic	One two-fold axis	Three axes of unequal length; two axes are not at right-angles; the third, the symmetry axis, is at right-angles to the plane containing the other two
Triclinic	Either a center of symmetry; or no symmetry	Three axes, all of unequal length, none at right-angles to the others
Hexagonal	One vertical six-fold axis	Four axes, three of equal length arranged in a horizontal plane; the fourth perpendicular to this plane and of different length from the other three
Trigonal	One vertical three-fold axis	As for hexagonal

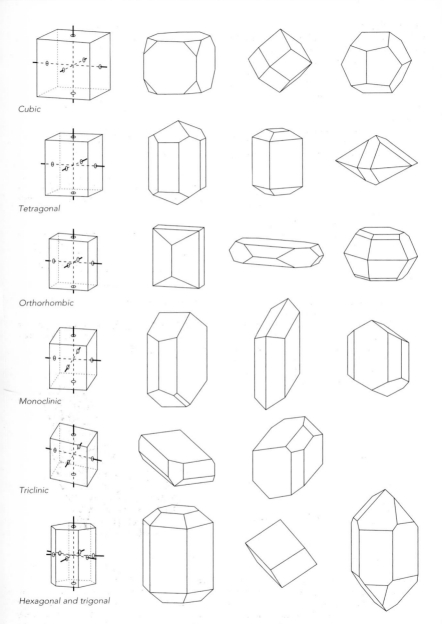

Cubic

Tetragonal

Orthorhombic

Monoclinic

Triclinic

Hexagonal and trigonal

▲ Fig. 5 Reference axes of the crystal systems and some examples of crystals belonging to each

Crystal form

It is useful when identifying minerals to determine to which crystal system they belong, but minerals that crystallize in the same crystal system, and even crystals of the same substance, can show remarkable differences in shape according to which crystal *form*, or combination of forms, is developed.

A crystal form comprises all the faces required by the symmetry. Some forms, such as the cube and octahedron (Fig. 6), totally

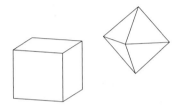

▲ *Fig. 6 Cube and octahedron*

enclose space and are called *closed forms* and can occur by themselves as crystals. Other forms, such as a pinacoid (a pair of parallel faces), or a prism (a form comprising three or more faces that meet in edges that are parallel) do not totally enclose space, and are called *open forms* (Fig. 7). Clearly they can occur only in *combination* with other forms, because crystals are solids.

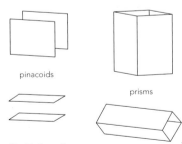

pinacoids

prisms

▲ *Fig. 7 Open forms*

Forms are often used to describe the appearance of minerals, for example spinel is octahedral, hornblende occurs as prismatic crystals (Fig. 8). A more detailed treatment of crystal shape will be found in the reference books.

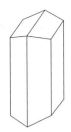

▲ *Fig. 8 Prismatic hornblende crystal*

The general aspect conferred on a mineral by the development of its faces is called the *habit*. Thus barite commonly forms crystals

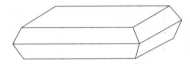

▲ *Fig. 9 Tabular habit*

of *tabular* habit (Fig. 9); and zeolites such as natrolite frequently have *acicular* (needle-like) habit (Fig. 10).

▲ *Fig. 10 Acicular habit*

Mineral aggregates

So far only single crystals have been discussed. Most minerals, however, occur as *aggregates* of crystals that rarely show perfect crystal shapes. The form of the aggregate, however, can be useful

▲ Fig. 11 Botryoidal form

▲ Fig. 12 Mammillary form

in identification. The *fibrous* zeolites have already been mentioned, and this adjective aptly describes their appearance. Sometimes crystals grow outward from a center, and the aggregate so formed is internally radiating, and outwardly may be rounded and nodular. The resulting form resembles a bunch of grapes and is called *botryoidal* (Fig. 11). Larger and more gently rounded shapes are said to be *mammillary* (Fig. 12). Minerals such as native copper often form distinctive branching and divergent forms to which the term *dendritic* is applied (Fig. 13), and crystals forming distinctly flat sheets are said to be *lamellar*. If the lamellae are very thin and can be readily separated, like the pages of a book, they are said to be *foliated* (Fig. 14). These and other examples are given in the mineral descriptions.

Physical properties

There is a close link between the structure of a mineral and its physical properties which are, accordingly, of considerable value in

identification. Some of the more useful physical properties are described below.

Density Defined strictly, density is mass per unit volume and is expressed in appropriate units, for example pounds per cubic inch. It is often used synonymously, though not strictly correctly, with *specific gravity*, which is the weight of the substance compared to the weight of an equal volume of water. Density depends on several factors including the kind of atoms in the structure and how closely they pack together. Other things being equal, the heavier the atoms or the more closely packed they are, the greater the density. Tridymite and quartz are both silica (SiO_2) but quartz, the closely packed form, has a specific gravity of 2.65 at room temperature, whereas tridymite, with a more open structure, has a specific gravity of 2.26 under the same conditions. Similarly, celestine and anglesite (sulfates of strontium and lead respectively) have the same structures, but the presence of the heavier lead atoms gives anglesite a specific gravity

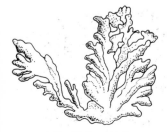

▲ Fig. 13 Dendritic form

▲ Fig. 14 Foliated lamellae

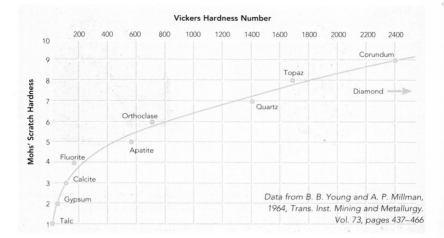

Vickers Hardness Number

Data from B. B. Young and A. P. Millman,
1964, Trans. Inst. Mining and Metallurgy.
Vol. 73, pages 437–466

of 6.32 compared with 3.97 for celestine. With practice, specific gravity can be roughly estimated by hand; methods of measurement are described in the books listed as "further reading".

Hardness Hardness is the resistance of a mineral to scratching or abrasion. F. Mohs, in 1812, arranged ten minerals in order of hardness, so that each will scratch those lower in the scale (depicted above).

It says much for Mohs' careful selection that this scale is still used as the standard for scratch hardness. Hardness is tested either by observing whether or not the minerals of Mohs' scale scratch the unknown mineral, or by observing whether objects of known hardness such as a knife blade or the fingernail will scratch the unknown. Minerals of hardness 1 feel soapy or greasy; the fingernail has a hardness of about $2\frac{1}{2}$; a steel pocket-knife blade has a hardness of about $5\frac{1}{2}$; and minerals of hardness 6 and over will scratch glass. Hardness is related to structure and to the strength of the chemical bonding; it is greater the smaller the atoms in the structure or the closer their packing. It should be appreciated that hardness is not the same as difficulty of breaking. A hard mineral may be brittle.

Cleavage and fracture If roughly handled, crystals will break. If the broken surface is irregular, the crystal possesses *fracture*, but if it breaks along a plane surface that is related to the structure, and parallel to a possible crystal face, then it has *cleavage*. Cleavage and fracture are expressions of the internal structure of the mineral. Cleavage occurs because of the variation in the strength of the bonds between different atoms, or planes of atoms. This is best illustrated by the layer silicates, of which mica is a familiar example. Chemical bonds are very strong within the silicon-oxygen layers, but the bonds between layers are weak, and so very little effort is needed to break them. Mica splits easily (cleaves) into thin sheets. Bond strength varies and so the degree of perfection of cleavage varies also. Mica, for example, has a *perfect* cleavage; less perfect cleavages are described as

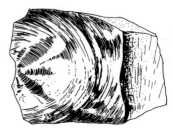

▲ Fig. 15 Conchoidal fracture

good, poor, or indistinct. Cleavages may develop in several directions within a crystal and their quality and direction may well be of diagnostic value. Fracture has no such structural control but may still be of use in mineral identification. Thus glass, which lacks an orderly arrangement of its atoms, breaks with a characteristic conchoidal fracture (Fig. 15), so called because the overall appearance of the fractured surface, with its concentric ridges, resembles a shell with its growth lines. Quartz, though crystalline, has such a uniformly bonded structure that it breaks with a conchoidal fracture similar to that of glass. Another important type of fracture gives a broken surface that resembles that of fractured wrought iron and is called hackly fracture.

Optical properties
Optical properties depend on the interaction of light with minerals.

Transparency A most obvious property is whether a mineral in hand specimen is transparent, translucent or opaque. This is a function of the structure of the mineral and the kind of bonds that bind atom to atom. It is a measure of the amount of light absorbed by the mineral, and the subject is treated fully in some of the recommended reading. Many minerals, which in the mass are opaque, become translucent in very thin fragments.

Reflection and refraction When light meets a translucent mineral at an oblique angle, part is reflected from its surface and part enters the crystal or is refracted into it. Refraction is of little diagnostic value in the field, but it is a most useful property of minerals when they are investigated in the laboratory.

Luster Luster is a property of the surface of a mineral. The nature of the reflecting surface gives rise to the different kinds of luster, and the amount of light reflected produces different intensities of luster. Luster, it should be remembered, is assessed independently of color. The main kinds of luster are described below.
Metallic luster is the luster of metals. It is produced by minerals which, like metals,

absorb light strongly and are opaque even in the thinnest slices. In addition to the native metals themselves, most sulfides have a metallic luster. Imperfect metallic luster is called submetallic. There are various kinds of non-metallic luster. Adamantine luster is the luster of diamond. Resinous luster is the luster of resin. This occurs in certain minerals with a yellow to brown color. Vitreous luster is the luster of broken glass. This is the luster most commonly displayed by minerals. Certain kinds of luster are caused by the quality of the reflecting surface. A greasy luster is often caused by minute irregularities in the surface which, if perfectly smooth, would give an adamantine or resinous luster. Pearly luster results from the reflection of light from a succession of parallel surfaces, such as cleavage planes, within a crystal. Silky luster is due to the presence of small parallel fibers, as in asbestos and some varieties of gypsum. Earthy luster is in effect a lack of luster produced by surfaces that scatter the light. It is worth remembering that luster may vary in different faces of a crystal. Heulandite, for example, shows a pearly luster on one pair of faces and vitreous luster on all the others.

Color Color in minerals is the result of the selective absorption of parts of the spectrum of white light, the observed color being due to those wavelengths of light that are least absorbed. There is no single cause of color in minerals. Sometimes it is a direct result of the presence in the structure of certain chemical elements; for example many copper minerals are blue or green. There are other more subtle reasons, however, and the reader is referred to the references for more information. It should be emphasized that, with experience, color is one of the most valuable of the diagnostic properties of minerals.

Streak Streak is the color of the powdered mineral. The most usual means of determining streak is to draw the mineral across a piece of white, unglazed porcelain, called a streak plate. Whereas the color of the mineral in the mass can often be very variable, the color of its streak is much less so. Streak is particularly valuable in the determination of opaque and colored minerals. It is of little diagnostic value in the silicates, most of which have a white streak, and are too hard to powder readily.

Fluorescence When certain minerals are irradiated with ultraviolet light, they emit light in the visible part of the spectrum and are said to fluoresce. Fluorite – from which the name of the phenomenon is derived – and many other minerals show this property. Although interesting and sometimes spectacular, fluorescence only occasionally ranks as an important diagnostic property, because its effects are so variable. Different specimens of a single mineral may fluoresce with several different colors, and even specimens from the same locality may vary considerably in their fluorescence.

Other properties Some minerals have distinctive magnetic, electrical, and radio-active properties of use in identification. They are mentioned, where appropriate, in the mineral descriptions, and full accounts are given in some of the books listed as further reading.

Chemistry of minerals

It is possible to write a chemical formula to express the composition of a mineral, and such formulae are used as a short way of expressing mineral chemistry. Atoms can conveniently be regarded as electrically neutral because the positive charge on the nucleus is balanced by the negative charges of the surrounding electrons. Atoms can, however, gain or lose one or more electrons and so become either negatively or positively charged, when they are called *ions*. Negatively charged ions are called *anions* and positive ions are called *cations*. A chemical compound can be regarded as being made up of two parts, a positively charged or cationic part and a negatively charged or anionic part. The resulting compound is electrically neutral because the two sets of charges are in balance. The positive part is usually a metal and is always the first part of a written chemical formula. The negative or anionic part of the formula can be either a non-metallic ion such as oxygen or sulfur or else a combination of several elements to form a negatively charged group such as carbonate (CO_3) or

sulfate (SO_4). The following table lists the chemical symbols of the elements referred to in this book.

Ag	Silver
Al	Aluminum
As	Arsenic
Au	Gold
B	Boron
Ba	Barium
Be	Beryllium
Bi	Bismuth
C	Carbon
Ca	Calcium
Cd	Cadmium
Ce	Cerium
Cl	Chlorine
Co	Cobalt
Cr	Chromium
Cu	Copper
F	Fluorine
Fe^{2+}, Fe^{3+}	Iron
H	Hydrogen
Hg	Mercury
K	Potassium
La	Lanthanum
Li	Lithium
Mg	Magnesium
Mn^{2+}, Mn^{3+}, Mn^{4+}	Manganese
Mo	Molybdenum
N	Nitrogen
Na	Sodium
Nb	Niobium
Ni	Nickel
O	Oxygen
P	Phosphorus
Pb	Lead
S	Sulfur
Sb	Antimony
Si	Silicon
Sn	Tin

Sr	Strontium
Ta	Tantalum
Th	Thorium
Ti	Titanium
U	Uranium
V	Vanadium
W	Tungsten
Y	Yttrium
Zn	Zinc
Zr	Zirconium

Some common anionic groups and their names are given below.

Al_2O_4 etc	Aluminate
As, As_2 etc	Arsenide
AsO_4 etc	Arsenate
BO_3, B_3O_4 etc	Borate
Cl, Cl_2 etc	Chloride
CO_3	Carbonate
CrO_4 etc	Chromate
F, F_2 etc	Fluoride
MoO_4 etc	Molybdate
N, N_2 etc	Nitride
NO_3	Nitrate
NbO_3 etc	Niobate
O, O_2 etc	Oxide
OH, $(OH)_2$ etc	Hydroxide
PO_4 etc	Phosphate
S, S_2 etc	Sulfide
SiO_4, Si_2O_7 etc	Silicate
SO_4	Sulfate
TaO_3 etc	Tantalate
TiO_3 etc	Titanate
UO_2 etc	Uranate
VO_4 etc	Vanadate
WO_4 etc	Tungstate

In chemical formulae the subscript numerals denote the numbers of atoms of the preceding element that are present in the formula unit. When referring to a chemical compound by name it is simply necessary to state, in turn, the cationic and then the anionic part that follows; for example $CaCO_3$ is calcium carbonate, FeS_2 is iron sulfide, CaF_2 is calcium fluoride, $(Mg,Fe)SiO_4$ is magnesium iron silicate, and so on. By contrast, $KAlSi_3O_8$ is potassium aluminum silicate, or better, potassium aluminosilicate; here there are two parts of the cationic group and they are emphasized in the way shown. Another example is $K_2(UO_2)_2(VO_4)_2.3H_2O$ which is called hydrated potassium uranylvanadate. Notice that water of crystallization (H_2O) is referred to by the adjective "hydrated". Atoms which can substitute one for the other in a mineral are written so (Mg,Fe).

Field occurrence

Nearly all rocks are composed of minerals, but fine specimens are rare and tend to occur in fissures and other cavities where the crystals have been unobstructed during their growth.

Many good specimens are obtained from mineral veins (Fig. 16). High-temperature fluids deposit minerals in cracks and fissures in rocks and many of these veins, often called hydrothermal veins, are worked as sources of ore. They frequently contain colorful specimens and good crystals, not only of the commercially valuable ore minerals, but also of the accompanying and economically valueless *gangue* minerals as well. It is not always necessary to examine or collect from the veins themselves – in many

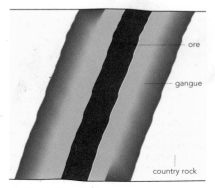

▲ Fig. 16 Mineral vein

instances it is dangerous or impossible to do so – for mining activity usually results in dumps of discarded material which, if carefully searched, will often yield good specimens. Good crystals can often be found lining cavities in rocks of virtually every kind, though particular minerals tend to occur in certain environments. Sometimes weathered-out cavity linings, called *geodes*, are lined with well shaped crystals, and many fine specimens of amethyst occur in such associations (Fig. 17). Pegmatites, which crystallize from relatively low-temperature, volatile-rich magma (fluids), are another source of good crystals and rare minerals that frequently grow to large sizes.

The largest specimens, however, are not always the most spectacular, and there is a growing interest in *micromounts* in which

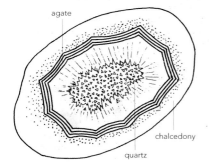

▲ *Fig. 17 Geode*

small crystals, or groups of crystals, are carefully mounted in a transparent plastic or glass-topped box in which they can be examined by using a lens or a microscope. These small crystals have a beauty of their own, and have the advantages of occupying a minimum of space and of being more perfectly formed than larger crystals.

Collectors will invariably find some specimens that are difficult to identify. They are urged to become acquainted with minerals that are displayed in many national and other museums. Time spent in this way will be amply repaid, not only in terms of identification of specimens, but also in becoming more deeply involved in the study of natural history.

Organization of the mineral descriptions in this book

The groups are described in the following order:

1 Native elements
2 Sulfides
3 Oxides and hydroxides
4 Halides
5 Carbonates
6 Nitrates and borates
7 Sulfates and chromates
8 Tungstates and molybdates
9 Phosphates, arsenates, and vanadates
10 Silicates

The silicates are such a large group that, although the primary classification is based on a chemical criterion, they are subdivided on a structural basis. This system has the advantage of grouping minerals with similar properties.

Gold Au

Crystal system Cubic. **Habit** Usually as disseminated grains, or dendritic forms: crystals rare but octahedral; occasionally as cubes or rhombdodecahedra. Irregular rounded masses are called nuggets. **Twinning** Common, on octahedron. **SG** 19.3 (less if alloyed with other metals). **Hardness** $2\frac{1}{2}$–3 **Cleavage** None. **Fracture** Hackly. **Color and transparency** Characteristic gold-yellow; lighter yellow when alloyed with silver: opaque except in thinnest sheets. **Streak** Gold-yellow. **Luster** Metallic. **Distinguishing features** Color, low hardness, insoluble in single acids. Gold may be mistaken for pyrite or chalcopyrite (fool's gold) but the color, low hardness and ductility of gold contrast with the greater hardness and brittle nature of the other two. **Alteration** None. **Occurrence** In small amounts in hydrothermal veins, often in association with quartz; and in alluvial deposits in which gold, by reason of its density, is separated from other minerals during weathering and transport to become concentrated in stream or other sediments, which may be loose and unconsolidated, or hardened into rock. Tiny grains of gold are often carried long distances by streams and can be recovered from gravel by panning, which entails washing away all but the heavy minerals, and searching these for flecks of gold. South African gold is obtained from consolidated alluvial deposits, notably gold-bearing quartz conglomerates.

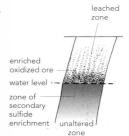

leached zone

enriched oxidized ore

water level

zone of secondary sulfide enrichment / unaltered zone

Mineral vein showing secondary sulfide enrichment

Gold on quartz

Gold nuggets

2 ins

Silver Ag

Crystal system Cubic. **Habit** Commonly as wiry or scaly forms; crystals rare. **SG** 10–11 **Hardness** $2\frac{1}{2}$–3 **Cleavage** None. **Fracture** Hackly. **Color and transparency** Silver-white, tarnishes quickly to a black color: opaque. **Streak** Silver-white. **Luster** Metallic. **Distinguishing features** Color, black tarnish, malleability, soluble in nitric acid. **Occurrence** In hydrothermal veins, or in small amounts in the oxidized zone of silver-bearing ore deposits.

Copper Cu

Crystal system Cubic. **Habit** Dendritic, branching forms; crystals usually cubic or rhombdodecahedral. **SG** 8.9 **Hardness** $2\frac{1}{2}$–3 **Cleavage** None. **Fracture** Hackly. **Color and transparency** Copper-red, deepening to dull brown with tarnish: opaque. **Streak** Metallic copper-red. **Luster** Metallic. **Distinguishing features** Color and ductility, readily soluble in nitric acid. **Occurrence** In basaltic lavas and in sandstones and conglomerates, in which it is secondary, having formed by reaction between copper-bearing solutions and other minerals, notably those of iron. Native copper, though widely distributed, occurs only in small amounts.

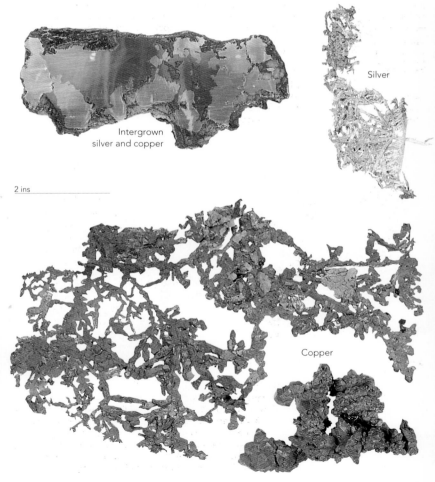

Intergrown
silver and copper

Silver

2 ins

Copper

17

Arsenic As
Crystal system Trigonal. **Habit** Crystals rare; usually massive as granular, botryoidal or stalactitic masses. **SG** 5.6–5.8 **Hardness** $3\frac{1}{2}$ **Cleavage** Basal, perfect. **Color and transparency** Light gray, tarnishes rapidly to dark gray: opaque. **Streak** Light gray. **Luster** Metallic. **Distinguishing features** Smells like garlic when heated or struck with a hammer. **Occurrence** In hydrothermal veins, usually in igneous and metamorphic rocks, and associated with silver, cobalt or nickel ores. The name arsenic is derived from a Greek word for "masculine", and dates from the time when metals were thought to be of different sexes.

Antimony Sb
Crystal system Trigonal. **Habit** Usually massive and reniform (kidney-shaped); sometimes lamellar; crystals rare. **Twinning** Common. **SG** 6.6–6.7 **Hardness** $3-3\frac{1}{2}$ **Cleavage** Basal, perfect; rhombohedral good. **Color and transparency** Very light gray: opaque. **Streak** Gray. **Luster** Metallic. **Occurrence** In hydrothermal veins, often associated with silver or arsenic. Accompanying minerals are stibnite, sphalerite, galena, and pyrite.

2 ins

Arsenic

Antimony

18

Bismuth Bi

Crystal system Trigonal. **Habit** Massive, granular or arborescent (tree-like or moss-like); crystals rare. **Twinning** Fairly common. **SG** 9.7–9.8 **Hardness** 2–2½ **Cleavage** Basal, perfect. **Color and transparency** Silver-white, becoming reddish with tarnish: opaque. **Streak** Silver-white, shiny. **Luster** Metallic. **Distinguishing features** Reddish silver color, perfect cleavage, melts readily at 520°F. **Occurrence** In hydrothermal veins, often in association with ores of gold, silver, tin, nickel, cobalt and lead.

Iron Fe **Kamacite, taenite** NiFe

Crystal system Cubic. **Habit** In grains and masses in terrestrial rocks. Kamacite and taenite (*see under meteorites*) differ in the amount of iron and nickel they contain, and are the major native metallic constituents of meteorites. **SG** 7.3–7.9 **Hardness** 4½ **Cleavage** Poor. **Fracture** Hackly. **Color and transparency** Steel-gray to black: opaque. **Luster** Metallic. **Distinguishing features** Strongly magnetic character, malleability. **Occurrence** Native iron is uncommon in terrestrial rocks, occurring mainly where volcanic rocks cut coal seams.

2 ins

Bismuth

Bismuth

Iron

Sulfur

Sulfur S

Crystal system Orthorhombic. **Habit** Crystals tabular or bipyramidal; also occurs as stalactitic or encrusting masses. **SG** 2.0–2.1 **Hardness** $1\frac{1}{2}$–$2\frac{1}{2}$ **Cleavage** None. **Fracture** Uneven, sometimes conchoidal. **Color and transparency** Bright yellow, sometimes brownish: transparent to translucent. **Streak** White. **Luster** Resinous. **Distinguishing features** Color, low hardness, low melting point (235°F), insoluble in water and dilute hydrochloric acid, soluble in carbon disulfide. **Occurrence** As encrusting masses produced by sublimation around volcanic vents and fumaroles; in sedimentary rocks, particularly limestones and those containing gypsum. Sulfur often occurs in the cap rock of salt domes in association with anhydrite, gypsum and calcite.

Cap rock containing sulfur above salt dome

Diamond C

Crystal system Cubic. **Habit** Commonly occurs as octahedral crystals frequently of flattened habit; more rarely as cubes, often with curved faces. **Twinning** Sometimes twinned on octahedron. **SG** 3.5 **Hardness** 10 **Cleavage** Octahedral, perfect. **Fracture** Conchoidal. **Color and transparency** Colorless: transparent. May be yellowish, brown, red, and even black. Opaque, finely granular diamond is called bort. **Streak** White. **Luster** Adamantine; uncut crystals look greasy. **Distinguishing features** Extreme hardness, octahedral cleavage. **Occurrence** Sporadically distributed in kimberlite and lamproitic rocks, forming pipe-like intrusions that have risen from great depth; also in alluvial deposits (mainly river and beach gravels) in which diamond is concentrated. Most diamonds came from alluvial deposits until the discovery of kimberlite pipes in South Africa in the mid-19th century. The name comes from the Greek word meaning "invincible" and alludes to the hardness and durability of diamond.

Diamond: octahedron

Graphite C

Crystal system Hexagonal. **Habit** Flat tabular crystals but more commonly massive, foliated or earthy. **SG** 2.1–2.3 **Hardness** 1–2 **Cleavage** Basal, perfect. **Color and transparency** Black: opaque. **Streak** Black. **Luster** Dull metallic. **Distinguishing features** Extreme softness, greasy feel, readily marks paper and soils the fingers. Distinguished from molybdenite by its black streak, lower specific gravity and color; molybdenite being bluish gray with gray to gray-green streak. **Occurrence** As disseminated flakes in metamorphic rocks derived from rocks having an appreciable carbon content. Graphite schists and limestones are fairly widely distributed. It occurs also as veins in igneous rocks and pegmatites. The name comes from the Greek word meaning "to write". Diamond and graphite have the same chemical composition and yet have quite different structures and physical properties. This phenomenon, in which a chemical substance can exist in two or more distinct forms which differ in structure and physical properties, is called *polymorphism*. There could hardly be a greater contrast in hardness than that between diamond and graphite.

Diamond: octahedron with curved faces

2 ins

Sulfur

Diamond in matrix

Bort

Diamonds

Graphite

Graphite

Argentite-acanthite (Silver glance) Ag₂S
Crystal system Cubic (argentite); monoclinic (acanthite). **Habit**
Crystals commonly cubic or octahedral, frequently occur as
groups of crystals in parallel alignment. Acanthite may crystallize
at low temperatures as pointed crystals. Also arborescent,
filiform (wiry), massive. **SG** 7.2–7.4 **Hardness** 2–2½ **Cleavage**
Cubic, poor. **Fracture** Subconchoidal. **Color and transparency**
Black: opaque. **Streak** Black, shiny. **Luster** Metallic. **Distinguishing
features** Color, sectility (can be cut with a knife, like lead).
Alteration Argentite is stable only above 360°F. Below this temperature Ag₂S has a monoclinic structure and is called acanthite.
The cubic forms shown are thus paramorphs of acanthite after
argentite. **Occurrence** In hydrothermal veins in association with
pyrargyrite, proustite and native silver. It also occurs as a weathering product of primary silver sulfides.

Bornite (Peacock ore, Erubescite) Cu₅FeS₄
Crystal system Tetragonal below about 450°F; cubic above.
Habit Crystals rough, pseudo-cubic and rhombdodecahedral;
usually massive. **Twinning** On octahedron. **SG** 5.0–5.1 **Hardness** 3
Cleavage None visible. **Fracture** Subconchoidal, uneven. **Color
and transparency** Reddish brown on fresh surface; tarnishes
to characteristic purplish iridescence: opaque. **Streak** Pale
gray-black. **Luster** Metallic. **Distinguishing features** Iridescent
colors, hence "peacock ore". Soluble in nitric acid. **Alteration** To

2 ins

Argentite-acanthite

Bornite

Covellite

Chalcosite

2 ins

chalcosite, covellite, cuprite, chrysocolla, malachite, and azurite.
Occurrence A common copper mineral, found in hydrothermal
veins in association with chalcopyrite and chalcosite. It occurs
also as a primary mineral in some igneous rocks and pegmatite
veins.

Covellite CuS
Crystal system Hexagonal. **Habit** Crystals tabular or platy; rare.
Usually as foliated masses or coatings. **SG** 4.6–4.8 **Hardness**
$1\frac{1}{2}$–2 **Cleavage** Basal, perfect. **Color and transparency** Indigo
blue, purplish iridescent tarnish: opaque. **Streak** Dark gray to
black. **Luster** Metallic. **Distinguishing features** Perfect cleavage
distinguishes covellite from bornite. Color distinguishes it
from chalcosite. **Occurrence** In hydrothermal veins as a primary
sulfide; more commonly in the zone of secondary enrichment in
association with chalcocite, bornite, and chalcopyrite.

Chalcosite

Chalcocite (Copper glance) Cu₂S
Crystal system Monoclinic; hexagonal above 220°F. **Habit** Prismatic or tabular crystals, rare; usually massive, or as powdery
coatings. **Twinning** Common, to give pseudo-hexagonal forms.
SG 5.5–5.8 **Hardness** $2\frac{1}{2}$–3 **Cleavage** Prismatic, indistinct.
Fracture Conchoidal. **Color and transparency** Dark lead-gray,
tarnishing to black: opaque. **Streak** Black. **Luster** Metallic.
Distinguishing features Black color, association with other
copper minerals. Soluble in nitric acid. **Alteration** To covellite,
malachite or azurite. **Occurrence** A widespread and valuable
copper ore, often associated with native copper or cuprite. Most
commonly found in the zone of secondary sulfide enrichment.

leached
zone

enriched
oxidized ore

water level

zone of
secondary
sulfide
enrichment

unaltered
zone

*Mineral vein showing
secondary sulfide
enrichment*

Sphalerite

2 ins

Sphalerite: combination of two tetrahedra and cube

Sphalerite (Zinc blende, Blende, Black Jack) (Zn,Fe^{2+})S
Crystal system Cubic. **Habit** Crystals commonly tetrahedral or rhombdodecahedral in combination with the cube; often distorted and with curved faces. Also granular, fibrous, botryoidal. **Twinning** Common, on octahedron, often repeated. **SG** 3.9–4.1 **Hardness** $3\frac{1}{2}$–4 **Cleavage** dodecahedral, perfect. **Fracture** Conchoidal. **Color and transparency** Commonly yellow, brown, black: transparent to translucent; sometimes appears opaque. **Streak** Brown to light yellow or white. **Luster** Resinous; nearly metallic in opaque specimens. **Distinguishing features** Sphalerite is very variable in color and can be difficult to recognize with certainty. The name in fact derives from a Greek word meaning "treacherous", because sphalerite is so easily mistaken for other minerals. The cleavage and resinous luster are reliable, and it is most commonly yellow to dark brown. **Alteration** To limonite; or to hemimorphite or smithsonite. **Occurrence** Sphalerite is the most common zinc mineral. It is frequently associated with galena in hydrothermal veins; it occurs also in limestones where ore bodies have formed by replacement, and where it is associated with pyrite, pyrrhotite, and magnetite.

Chalcopyrite

Chalcopyrite (Copper pyrites) CuFeS$_2$
Crystal system Tetragonal. **Habit** Crystals appear tetrahedral; usually massive. **Twinning** Several kinds, giving rise to interpenetration twins and twins that resemble spinel twins. **SG** 4.1–4.3 **Hardness** $3\frac{1}{2}$–4 **Cleavage** Very poor. **Fracture** Conchoidal to uneven. **Color and transparency** Brass-yellow, often with slightly iridescent tarnish: opaque. **Streak** Greenish black. **Luster** Metallic. **Distinguishing features** Distinguished from pyrite by its deeper yellow color, tarnish and inferior hardness; and from gold by its brittle nature and greater hardness. Soluble in nitric acid. **Alteration** To chalcosite, covellite, chrysocolla, and malachite. **Occurrence** Chalcopyrite is the most common copper

2 ins

Chalcopyrite

mineral, and an important copper ore. It occurs as a primary mineral in igneous rocks and in hydrothermal vein deposits in association with pyrite, pyrrhotite, cassiterite, sphalerite, galena, and gangue minerals such as quartz, calcite, and dolomite. It is an important mineral in "porphyry copper" deposits where it is disseminated with bornite and pyrite in veinlets in igneous intrusions of quartz diorite or diorite porphyry. It occurs also in pegmatites, crystalline schists, and in contact metamorphic deposits.

Wurtzite ZnS

Crystal system Hexagonal. **Habit** Pyramidal crystals; also radiating, fibrous, massive. **SG** 4.0–4.1 **Hardness** $3\frac{1}{2}$–4 **Cleavage** Prismatic, distinct; basal, imperfect. **Color** Brownish black. **Streak** Brown. **Luster** Resinous. **Occurrence** Rare; in sulfide ores. Wurtzite is a polymorph of ZnS, and is the form stable at high temperatures. It is named after C. A. Wurtz, a French chemist.

Wurtzite

Galena

2 ins

Galena: cube

Galena: cube and octahedron

Galena PbS

Crystal system Cubic. **Habit** Crystals often of cube/octahedral, or octahedral habit; sometimes as cubes. Also massive or granular. **Twinning** Penetration or contact twins on octahedron. **SG** 7.4–7.6 **Hardness** $2\frac{1}{2}$ **Cleavage** Cubic, perfect. **Color and transparency** Lead-gray: opaque. **Streak** Lead-gray. **Luster** Metallic. **Distinguishing features** Color, metallic luster, perfect cubic cleavage, high specific gravity. **Alteration** Oxidizes readily to anglesite, cerussite, pyromorphite, or mimetite. **Occurrence** Galena is very widely distributed and the most important lead ore. It occurs, following the bedding of sedimentary rocks, as hydrothermal veins and in pegmatites, and as replacement bodies in limestone and dolomitic rocks. In hydrothermal veins it is commonly associated with sphalerite, pyrite, chalcopyrite, tetrahedrite, and bournonite, and with gangue minerals such as quartz, calcite, dolomite, barite, and fluorite. In high-temperature veins it is associated with such minerals as garnet, feldspar, diopside, rhodonite and biotite. Replacement deposits occur in limestones which have sometimes been dolomitized. The name comes from the Latin word for "lead ore".

Pyrrhotite (Magnetic pyrites, Pyrrhotine) FeS (varies to $Fe_{0.8}S$)
Crystal system Monoclinic; hexagonal at high temperatures.
Habit Usually massive granular; crystals rare, platy or tabular.
Twinning Rare. **SG** 4.6–4.7 **Hardness** $3\frac{1}{2}$–$4\frac{1}{2}$ **Cleavage** Basal parting. **Fracture** Subconchoidal to uneven. **Color and transparency**
Bronze-yellow, darkening with exposure to more reddish bronze:
opaque. **Streak** Grayish black. **Luster** Metallic. **Distinguishing
features** Reddish bronze color, magnetism. Apart from magnetite it is the only common mineral that is noticeably magnetic.
Distinguished from chalcopyrite by color and magnetism; from
pyrite by color and inferior hardness. **Occurrence** Pyrrhotite
occurs in igneous rocks such as gabbro as disseminated grains,
commonly in association with minerals such as chalcopyrite,
pentlandite, and pyrite. It is found also in contact metamorphic
deposits, in veins, and in pegmatites. It occurs (troilite) in
iron meteorites. The name comes from a Greek word meaning
"reddish".

2 ins

Pyrrhotite

27

Nickeline (Niccolite, Kupfernickel) NiAs
Crystal system Hexagonal. **Habit** Usually massive in reniform or columnar aggregates; crystals rare. **SG** 7.8 **Hardness** $5-5\frac{1}{2}$ **Cleavage** None. **Fracture** Uneven. **Color and transparency** Pale copper-red: opaque. **Streak** Pale brownish black. **Luster** Metallic. **Distinguishing features** Color. **Alteration** Alters to pale green annabergite (nickel bloom). **Occurrence** It occurs in igneous rocks such as norite and gabbro with pyrrhotite, chalcopyrite and nickel sulfides, and in hydrothermal veins with silver, silver-arsenic, and cobalt minerals.

Greenockite CdS
Crystal system Hexagonal. **Habit** Usually as a powdery coating; rarely as distinct crystals. **SG** 4.9–5.0 **Hardness** $3-3\frac{1}{2}$ **Cleavage** Prismatic, distinct; basal, imperfect. **Fracture** Conchoidal. **Color and transparency** Orange-yellow: nearly transparent. **Streak** Reddish yellow. **Luster** Adamantine to resinous. **Distinguishing features** Yellow color, powdery form, soluble in hydrochloric acid yielding hydrogen sulfide. **Occurrence** Occurs as a yellow coating with zinc minerals such as sphalerite. It is named in honor of Lord Greenock.

Greenockite

Cinnabar HgS
Crystal system Trigonal. **Habit** Crystals rhombohedral or thick tabular, sometimes short prismatic or acicular. Also granular, massive. **Twinning** Common, with basal pinacoid as twin plane. **SG** 8.0–8.2 **Hardness** $2-2\frac{1}{2}$ **Cleavage** Prismatic, perfect. **Fracture** Uneven. **Color and transparency** Scarlet-red to brownish red: transparent to translucent, occasionally nearly opaque. **Streak** Vermilion. Powdered cinnabar was used as a pigment. **Luster** Adamantine; near metallic when opaque. **Distinguishing features** Red color and streak, high specific gravity, perfect cleavage. **Alteration** Sometimes alters to calomel (mercurous chloride). **Occurrence** Cinnabar is the commonest of the mercury minerals, and is the only important ore of the metal. It occurs in fractures in sedimentary rocks in areas of recent volcanic activity and around hot springs. It is associated with pyrite, stibnite and realgar, and with gangue minerals such as chalcedony, quartz, calcite, and barite.

Cinnabar: thick tabular habit

Millerite NiS
Crystal system Trigonal. **Habit** Crystals usually slender and acicular, often in radiating groups. **SG** 5.2–5.6 **Hardness** $3-3\frac{1}{2}$ **Cleavage** Rhombohedral, perfect. **Fracture** Uneven. **Color and transparency** Brass-yellow: opaque. **Streak** Greenish black. **Luster** Metallic. **Distinguishing features** Color, slender acicular crystal form. Individual acicular crystals are elastic. **Occurrence** Commonly occurs as tufts of radiating fibres in cavities and as a replacement of other nickel minerals. It also occurs in veins carrying nickel minerals and other sulfides, and around some volcanoes as a sublimation product. It is named after W. H. Miller (1801–80), a British mineralogist.

Greenockite

Cinnabar

Nickeline

2 ins

Millerite

Cinnabar

Realgar

Realgar AsS

Crystal system Monoclinic. **Habit** Short prismatic crystals, striated parallel to their length. Also granular, compact. **SG** 3.5 **Hardness** 1½–2 **Cleavage** Pinacoidal, good. **Fracture** Conchoidal. **Color and transparency** Red to orange-yellow: transparent to translucent. **Streak** Orange-red. **Luster** Resinous. **Distinguishing features** Red color, low hardness, resinous luster, association with orpiment. **Alteration** On long exposure to light it breaks down to a yellow powder. **Occurrence** Commonly as a minor constituent of hydrothermal veins carrying arsenic minerals; also as a hot-spring deposit, and in limestones and dolostones.

Realgar

2 ins

Orpiment As₂S₃

Crystal system Monoclinic. **Habit** Usually as foliated or columnar masses; crystals small and rare. **SG** 3.4–3.5 **Hardness** 1½–2 **Cleavage** One perfect cleavage. **Color and transparency** Lemon-yellow to brownish or reddish yellow: transparent to translucent. **Streak** Pale yellow. **Luster** Pearly on cleavage surface; elsewhere resinous. **Distinguishing features** Yellow color, perfect cleavage, pearly luster on cleavage surface. **Occurrence** Orpiment often accompanies realgar as a low-temperature mineral in veins and hot-spring deposits.

Stibnite (Antimonite, Antimony glance) Sb₂S₃

Crystal system Orthorhombic. **Habit** Prismatic crystals, striated parallel to their length, sometimes curved. Acicular crystals commonly as radiating groups or random aggregates. Occasionally granular, massive. **SG** 4.5–4.6 **Hardness** 2 **Cleavage** One perfect cleavage parallel to length of crystals. **Fracture** Subconchoidal. **Color and transparency** Lead-gray, sometimes tarnished and iridescent: opaque. **Streak** Lead-gray. **Luster** Metallic. **Distinguishing features** Habit, perfect cleavage, low hardness. Melts readily, even in a match flame. Will ignite a safety match drawn across it. **Occurrence** Stibnite, the most common antimony mineral, is most commonly found with quartz in hydrothermal veins, as replacement bodies in limestone, and in hot-spring deposits. Often associated with realgar, orpiment, galena, pyrite, and cinnabar.

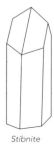

Stibnite

Jamesonite

Stibnite

Bismuthinite

Orpiment

2 ins

Jamesonite $Pb_4FeSb_6S_{14}$

Crystal system Monoclinic. **Habit** Acicular crystals; also fibrous, massive. **SG** 5.5–6.0 **Hardness** 2–3 **Cleavage** Basal, perfect. **Fracture** Uneven to conchoidal. **Color and transparency** Dark lead-gray: opaque. **Streak** Grayish black. **Luster** Metallic. **Distinguishing features** Distinguished from stibnite by lack of cleavage parallel to length of crystals. **Occurrence** Jamesonite occurs in veins with galena, sphalerite, pyrite, stibnite, etc.

Bismuthinite (Bismuth glance) Bi_2S_3

Crystal system Orthorhombic. **Habit** Usually massive, fibrous; rarely as acicular crystals. **SG** 6.8 **Hardness** 2 **Cleavage** One perfect cleavage. **Color and transparency** Light lead-gray: opaque. **Streak** Light lead-gray. **Luster** Metallic. **Distinguishing features** Similar to stibnite but sectile and less flexible. **Occurrence** In igneous rocks in association with such minerals as magnetite, pyrite, chalcopyrite, sphalerite, and galena, and with tin and tungsten ores.

2 ins

Pyrite

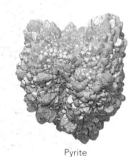

Pyrite

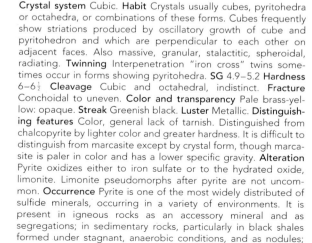

Pyrite

Pyrite: striated cube

Pyrite: pyritohedron

Pyrite (Iron pyrites) FeS_2

Crystal system Cubic. **Habit** Crystals usually cubes, pyritohedra or octahedra, or combinations of these forms. Cubes frequently show striations produced by oscillatory growth of cube and pyritohedron and which are perpendicular to each other on adjacent faces. Also massive, granular, stalactitic, spheroidal, radiating. **Twinning** Interpenetration "iron cross" twins sometimes occur in forms showing pyritohedra. **SG** 4.9–5.2 **Hardness** $6-6\frac{1}{2}$ **Cleavage** Cubic and octahedral, indistinct. **Fracture** Conchoidal to uneven. **Color and transparency** Pale brass-yellow: opaque. **Streak** Greenish black. **Luster** Metallic. **Distinguishing features** Color, general lack of tarnish. Distinguished from chalcopyrite by lighter color and greater hardness. It is difficult to distinguish from marcasite except by crystal form, though marcasite is paler in color and has a lower specific gravity. **Alteration** Pyrite oxidizes either to iron sulfate or to the hydrated oxide, limonite. Limonite pseudomorphs after pyrite are not uncommon. **Occurrence** Pyrite is one of the most widely distributed of sulfide minerals, occurring in a variety of environments. It is present in igneous rocks as an accessory mineral and as segregations; in sedimentary rocks, particularly in black shales formed under stagnant, anaerobic conditions, and as nodules; and in metamorphic rocks, notably in slates when it frequently

2 ins

Pyrite

Marcasite

forms well shaped cubic crystals. It is a common mineral in hydrothermal sulfide veins, in replacement deposits. Fossils are often replaced by pyrite. The name comes from the Greek word for "fire" and alludes to the sparks given off when the mineral is struck sharply.

Marcasite FeS$_2$

Crystal system Orthorhombic. **Habit** Crystals commonly tabular; also massive stalactitic or as radiating fibers. **Twinning** Common; producing spear-shaped forms or "cockscomb aggregates". **SG** 4.8–4.9 **Hardness** 6–6$\frac{1}{2}$ **Cleavage** Prismatic, poor. **Fracture** Uneven. **Color and transparency** Pale bronze-yellow: opaque. **Streak** Grayish black. **Luster** Metallic. **Distinguishing features** Very similar to pyrite but has lower specific gravity, paler color and distinctive spear-shaped forms. **Alteration** Like pyrite, it oxidizes easily to ferrous sulfate or limonite. It may also alter to pyrite. **Occurrence** Marcasite is deposited at low temperatures (less than 840°F) in hydrothermal veins containing zinc and lead ores. It is found in near-surface deposits, most commonly in sedimentary rocks such as limestone, especially chalk, and clay, as single crystals, concretions or as a replacement of fossils. The name comes from an Arabic word once used for pyrite.

Pyrite: "iron-cross" twin

Marcasite: spear-shaped twin

33

2 ins

Cobaltite

Arsenopyrite

Arsenopyrite

Arsenopyrite

Arsenopyrite (Mispickel) FeAsS

Crystal system Monoclinic, pseudo-orthorhombic. **Habit** Prismatic crystals common, faces often striated. Columnar crystals have a rhombic cross-section. Also granular, columnar, massive. **Twinning** Common, on prism. **SG** 5.9–6.2 **Hardness** $5\frac{1}{2}$–6 **Cleavage** Prismatic, indistinct. **Fracture** Uneven. **Color and transparency** Silver gray-white, often with a brownish tarnish: opaque. **Streak** Dark grayish black. **Luster** Metallic. **Distinguishing features** Silver-white color, crystal form. **Occurrence** Arsenopyrite forms under high to moderate temperature conditions and so it frequently accompanies gold, and ores of tin, tungsten, and silver, together with sphalerite, pyrite, chalcopyrite, galena, and quartz. Besides occurring in mineral veins, it is found disseminated in limestones, dolostones, gneisses, and pegmatites.

Cobaltite CoAsS

Crystal system Orthorhombic; pseudo-cubic. **Habit** Crystallizes as cubes, pyritohedra, or combinations of these forms. Also granular, massive. **SG** 6.0–6.3 **Hardness** $5\frac{1}{2}$ **Cleavage** Pseudo-cubic, perfect. **Fracture** Uneven. **Color and transparency** Silver-white to gray with reddish tinge: opaque. **Streak** Gray-black. **Luster** Metallic. **Distinguishing features** Cleavage, silver-white color and inferior hardness distinguish it from pyrite. **Occurrence** Cobaltite occurs in high-temperature hydrothermal veins together with skutterudite, arsenopyrite, and nickeline. It occurs as disseminated grains in metasomatic contact deposits.

Molybdenite MoS_2

Crystal system Hexagonal. **Habit** Crystals hexagonal, often tabular. Commonly occurs as foliated or scaly masses; also granular, massive. **SG** 4.6–4.8 **Hardness** $1-1\frac{1}{2}$ **Cleavage** Basal, perfect; laminae flexible but not elastic. **Color and transparency** Pale bluish lead-gray: opaque. **Streak** Greenish gray; bluish gray on paper. **Luster** Metallic. **Distinguishing features** Similar hardness to graphite but greater specific gravity. The lighter, bluish tinge is distinctive, contrasting with the lead-gray of graphite. Greasy feel. **Occurrence** Molybdenite is widely distributed but never in large quantities. It is an accessory mineral in granites and occurs in associated pegmatites and quartz veins. It occurs also in contact metamorphic deposits with garnet, pyroxenes, scheelite, pyrite and tourmaline, and in veins together with scheelite, wolframite, cassiterite, and fluorite. Molybdenite is an ore of molybdenum, and the name comes from the Greek word for "lead".

Molybdenite
with quartz

2 ins

Molybdenite

Chloanthite

Smaltite

Skutterudite

2 ins

Skutterudite: octahedral habit

Skutterudite $CoAs_{2-3}$ – smaltite $(Co,Ni)As_2$ – chloanthite $NiAs_2$ series

Crystal system Cubic. **Habit** Often massive, granular; crystals cubic, octahedral, or combinations of these forms; pyritohedra may also be present. **SG** 5.7–6.9 **Hardness** $5\frac{1}{2}$–6 **Cleavage** Cubic and octahedral, indistinct. **Fracture** Uneven. **Color and transparency** Tin-white, steel-gray when massive; iridescent or grayish tarnish: opaque. **Streak** Grayish black. **Luster** Metallic. **Distinguishing features** The three names are applied to a series ranging from skutterudite (cobalt arsenide), through smaltite, which is essentially cobalt-nickel arsenide, to chloanthite, which is nickel arsenide with excess arsenic. It is very difficult to distinguish the skutterudite minerals from arsenopyrite without chemical tests. **Occurrence** In veins, accompanying other cobalt and nickel minerals such as cobaltite and nickeline. Native silver, arsenopyrite, and calcite are also associated minerals. Skutterudite is the recognized name for the series as a whole; the other two are varieties of skutterudite. The name comes from the Norwegian locality, Skutterud.

Pyrargyrite Ag_3SbS_3

Crystal system Trigonal. **Habit** Crystals prismatic; also as massive aggregates. **Twinning** Common. **SG** 5.8 **Hardness** $2\frac{1}{2}$ **Cleavage**

Pyrargyrite

Proustite

2 ins

Rhombohedral, distinct. **Fracture** Uneven. **Color and transparency** Black when opaque, but deep red by transmitted light. Darkens on exposure to light. Translucent to nearly opaque, transparent in thin fragments. **Streak** Purplish red. **Luster** Adamantine; nearly metallic when opaque. **Distinguishing features** Deep red color and streak; deeper in color and less translucent than proustite. **Occurrence** Pyrargyrite and proustite are called ruby silver ores. They occur typically in low-temperature silver veins together with silver, argentite, tetrahedrite, galena, sphalerite, etc. The name comes from two Greek words meaning "fire" and "silver", in reference to its red color and its composition.

Proustite Ag_3AsS_3

Crystal system Trigonal. **Habit** Crystals prismatic, rhombohedral or scalenohedral. Also massive, compact. **Twinning** Common. **SG** 5.6 **Hardness** $2-2\frac{1}{2}$ **Cleavage** Rhombohedral, distinct. **Fracture** Uneven. **Color and transparency** Scarlet, darkens on exposure to light: translucent. **Streak** Scarlet-vermilion. **Luster** Adamantine. **Distinguishing features** Color, vermilion streak; lighter in color than pyrargyrite. **Occurrence** Proustite occurs together with pyrargyrite in silver veins. It is less common than pyrargyrite and, like it, is an ore of silver. Proustite is named after J. L. Proust (1755–1826), a French chemist.

Proustite: combination of prism, scaleno hedron and two rhombohedra

37

Tetrahedrite: tetrahedron

Tetrahedrite: modified tetrahedron

Tetrahedrite-tennantite series (Fahlerz) (Cu,Fe)$_{12}$(Sb,As)$_4$S$_{13}$
Crystal system Cubic. **Habit** Crystals commonly tetrahedral; also massive, granular, compact. **Twinning** Contact or penetration twins on tetrahedron. **SG** 4.6–5.1 **Hardness** 3–4$\frac{1}{2}$ **Cleavage** None. **Fracture** Subconchoidal to uneven. **Color and transparency** Dark gray to black: opaque. **Streak** Dark gray or brown to black. **Lustre** Metallic. **Distinguishing features** Tetrahedral form, gray-black color. There is a continuous series between tetrahedrite and tennantite as arsenic substitutes for antimony. **Alteration** Oxidizes to minerals such as malachite and azurite. **Occurrence** Commonly found in hydrothermal veins together with silver, copper, lead, and zinc minerals; also in igneous contact metamorphic deposits. Tennantite is less widespread in occurrence than tetrahedrite and occurs in metasomatic deposits in limestone, whereas tetrahedrite is more commonly found in lead-silver veins.

Tennantite

Enargite

Tetrahedrite

2 ins

Enargite Cu$_3$AsS$_4$
Crystal system Orthorhombic. **Habit** Crystals usually small, tabular or prismatic; often massive, granular, bladed or columnar. **Twinning** Sometimes produces star-shaped forms comprising three individuals. **SG** 4.4 **Hardness** 3 **Cleavage** Prismatic, perfect; pinacoidal, distinct. **Fracture** Uneven. **Color and transparency** Dark gray to black: opaque. **Streak** Black. **Luster** Metallic. **Distinguishing features** Color, two cleavages. It fuses easily and will melt in a match flame. **Occurrence** Enargite is not a common mineral. It occurs in low temperature, near-surface deposits in association with chalcosite, bornite, covellite, pyrite, sphalerite, tetrahedrite, barite, and quartz.

Enargite

Bournonite

Boulangerite

2 ins

Bournonite (Wheel-ore) $PbCuSbS_3$

Crystal system Orthorhombic. Habit Crystals tabular; also massive, granular, compact. Twinning Very common; repeated twinning produces crystals reminiscent of cog wheels. SG 5.7–5.9 Hardness $2\frac{1}{2}$–3 Cleavage None, or poor. Fracture Uneven. Color and transparency Gray to black: opaque. Streak Gray to black. Luster Metallic. Distinguishing features Habit of twinned crystals, also high specific gravity. Fuses easily. Alteration Occasionally alters to cerussite, malachite or azurite. Occurrence In hydrothermal veins accompanying such minerals as galena, chalcopyrite, tetrahedrite, stibnite, sphalerite. It has been used as an ore of copper, lead and antimony. Bournonite is named in honor of Count J. L. de Bourn'on (1751–1825), a French mineralogist.

Bournonite: cog-wheel shaped twins

Boulangerite $Pb_5Sb_4S_{11}$

Crystal system Monoclinic. Habit Crystals usually elongated prismatic; also as fibrous or plumose masses. SG 5.7–6.3 Hardness $2\frac{1}{2}$–3 Cleavage One good cleavage. Color and transparency Lead-gray; sometimes with yellow spots caused by oxidation: opaque. Streak Red-brown. Luster Metallic. Distinguishing features Similar in appearance to stibnite and jamesonite and very difficult to distinguish from them. Occurrence Commonly occurs in veins together with stibnite, galena, sphalerite, pyrite, and with quartz, dolomite, and calcite as gangue minerals.

Cuprite: octahedron

Cuprite (Red copper ore) Cu_2O
Crystal system Cubic. **Habit** Crystals usually octahedral, sometimes cubic or rhombdodecahedral, or combinations of these forms; also acicular. Also massive, granular. **SG** 5.8–6.1 **Hardness** $3\frac{1}{2}$–4 **Cleavage** None. **Fracture** Uneven. **Color and transparency** Red, though sometimes so dark as to be nearly black: subtranslucent; subtransparent when very thin. **Streak** Brownish red. **Luster** Adamantine or submetallic. **Distinguishing features** Cuprite is similar in color to hematite and cinnabar, but it is softer than hematite and harder than cinnabar; it differs also in color of the streak. **Alteration** Malachite pseudomorphs after cuprite are not uncommon. **Occurrence** Cuprite is usually formed as a secondary mineral in the oxidized zone of copper deposits, and is commonly accompanied by malachite, azurite and chalcosite. Fine, hair-like crystals of cuprite are called chalcotrichite. It is an ore of copper, and the name derives from the Latin *cuprum* for copper.

Tungstite $WO_3.H_2O$
Crystal system Orthorhombic. **Habit** Powdery or earthy coatings. **Cleavage** Basal, perfect. **Color** Yellow or yellowish green. **Luster** Earthy. **Distinguishing features** Yellowish color and association with other tungsten minerals. **Occurrence** Tungstite is a secondary mineral found in association with wolframite.

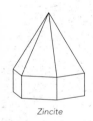

Zincite

Zincite $(Zn,Mn^{2+})O$
Crystal system Hexagonal. **Habit** Crystals very rare; usually massive, foliated, granular. **SG** 5.4–5.7 **Hardness** 4–$4\frac{1}{2}$ **Cleavage** Prismatic, distinct; basal parting. **Fracture** Subconchoidal. **Color and transparency** Deep red to orange-yellow: translucent. **Streak** Orange-yellow. **Luster** Subadamantine. **Distinguishing features** Red color, orange-yellow streak. Dissolves in hydrochloric acid. **Occurrence** Zincite is a rare mineral. In the few places where it does occur, notably at Franklin, New Jersey, USA, it is associated with franklinite and willemite in a contact metamorphic deposit.

Franklinite $(Zn,Mn^{2+},Fe^{2+})(Fe^{3+},Mn^{3+})_2O_4$
Crystal system Cubic. **Habit** Crystals octahedral; also massive, granular. **SG** 5.0–5.2 **Hardness** $5\frac{1}{2}$–$6\frac{1}{2}$ **Cleavage** None; octahedral parting. **Fracture** Uneven. **Color and transparency** Black: opaque. **Streak** Reddish brown to dark brown. **Luster** Metallic. **Distinguishing features** Franklinite, a member of the spinel group, most closely resembles magnetite, but is only slightly magnetic and has a dark brown streak. **Occurrence** Franklinite, zincite and willemite occur together in the zinc deposits of Franklin, New Jersey, USA. The deposits are associated with crystalline limestone and are probably of metasomatic origin. The deposit is worked as an ore of zinc and manganese, and the name of the mineral is taken from the locality.

Cuprite

Tungstite

Zincite

Zincite and franklinite

Franklinite

2 ins

Spinel: octahedron

Spinel: twinned octahedron

Spinel MgAl₂O₄

Crystal system Cubic. **Habit** Crystals usually octahedral; also massive. **Twinning** Common on octahedron, giving "spinel twins". **SG** 3.5–4.1 **Hardness** $7\frac{1}{2}$–8 **Cleavage** None; octahedral parting. **Fracture** Conchoidal. **Color and transparency** Very variable; commonly red (ruby spinel), but also blue, green, brown, black or colorless: transparent to nearly opaque, usually translucent. **Streak** White, but can be gray or even brown. **Luster** Vitreous. **Distinguishing features** Octahedral form, twinning, hardness. Spinel is the name of a series, rather than of a specific mineral. Iron, zinc, chromium and manganese atoms can substitute for magnesium in the structure and so give rise to variations in color and other physical properties. **Occurrence** Spinel occurs as an accessory mineral in igneous rocks such as gabbro. It also occurs in contact metamorphosed impure dolomitic limestones in association with phlogopite, graphite and chondrodite, and in aluminous metamorphic rocks. Gem-quality spinels occur in contact altered limestones and in alluvial gravels derived from them in Burma (Myanmar), Sri Lanka and India. Its hardness and resistance to weathering result in spinel being found as rolled pebbles in river and beach sands. The origin of the name is unknown.

Spinel

Spinel (Ceylonite)

2 ins

Magnetite

2 ins

Magnetite $Fe^{2+}Fe^{3+}_2O_4$
Crystal system Cubic. **Habit** Crystals most commonly octa-
hedra, also rhombdodecahedra; also massive, granular. **Twinning**
Common on octahedron. **SG** 5.2 **Hardness** $5\frac{1}{2}-6\frac{1}{2}$ **Cleavage**
None; octahedral parting. **Fracture** Subconchoidal to uneven.
Color and transparency Black: opaque. **Streak** Black. **Luster**
Metallic, shining; to submetallic, dull. **Distinguishing features**
Color and streak, strongly magnetic character. **Occurrence** Magnetite is very widely distributed in several environments. It is a
common accessory mineral in igneous rocks, and its relatively
high specific gravity sometimes results in it accumulating in them
to form deposits of economic value. It is a common mineral in
contact and regionally metamorphosed rocks and occurs in
high-temperature mineral veins. It often occurs in association
with corundum in emery deposits. Magnetite, by reason of its
strongly magnetic properties, has attracted attention since early
times. Specimens possessing polarity (lodestone) act as compass
needles when free to swing. Magnetite is an important iron ore.
There is some dispute as to the origin of the name. It most probably derives from its magnetic property, but it may derive from
Magnesia, a locality in Asia Minor, or be associated with the fable
of Magnes, a shepherd who is alleged to have discovered the
mineral on Mount Ida when the nails in his shoes and the iron
ferrule of his staff adhered to the ground.

43

Hematite

2 ins

Chromite $Fe^{2+}Cr_2O_4$ (often with some Mg and Al)
Crystal system Cubic. **Habit** Crystals rare, octahedral. Usually massive, granular. **SG** 4.1–5.1 **Hardness** $5\frac{1}{2}$ **Cleavage** None. **Fracture** Uneven. **Color and transparency** Black to brownish black: opaque, translucent in thin fragments. **Streak** Dark brown. **Luster** Metallic to submetallic. **Distinguishing features** Brown streak and weakly magnetic character distinguish chromite from magnetite. **Occurrence** Chromite is an accessory mineral in igneous rocks such as peridotite and in serpentinite. It may be concentrated into layers or lenses in sufficient quantity to be worked as an ore and, in fact, chromite is the only ore of chromium. Its durability sometimes results in chromite being concentrated in alluvial sands and gravels. It is an extremely refractory mineral and chromite bricks are used to line blast furnaces.

Hematite Fe$_2$O$_3$

Crystal system Trigonal. **Habit** Crystals tabular or rhombohedral, sometimes with curved and striated rhombohedral faces. Also columnar, laminated or massive, often in striking mammillary or botryoidal forms. **Twinning** Penetration twins on basal pinacoid. **SG** 4.9–5.3 **Hardness** 5–6 **Cleavage** None. **Fracture** Uneven, brittle. **Color and transparency** Steel-gray to black, sometimes iridescent. Massive compact varieties vary from dull to bright red. Opaque, except in very thin flakes. **Streak** Red to reddish brown. **Luster** Metallic, sometimes dull. **Distinguishing features** Red streak, hardness. **Occurrence** Hematite is the most important ore of iron and is widely distributed. It occurs as an accessory mineral in igneous rocks and in hydrothermal veins. It is of widespread occurrence in sedimentary rocks in which it may be of primary origin, often occurring as ooliths or as a cementing material; or as a secondary mineral, precipitated from iron-bearing percolating waters and replacing other minerals. Bedded iron formations of Precambrian age have afforded huge quantities of hematite in North America and elsewhere. In addition to its use as an iron ore, hematite is used as a pigment and as a polishing powder. The name is derived from the Greek word for "blood" and is descriptive of the color of the powdered mineral.

Hematite

Hematite: mammillary form

Chromite

2 ins

Ilmenite $Fe^{2+}TiO_3$

Crystal system Trigonal. **Habit** Crystals thick tabular; often massive, compact. **Twinning** Common, on basal pinacoid. **SG** 4.5–5.0 **Hardness** 5–6 **Cleavage** None; basal parting. **Fracture** Conchoidal. **Color and transparency** Black: opaque. **Streak** Black to brownish red. **Luster** Submetallic to metallic. **Distinguishing features** Distinguished from magnetite by non-magnetic character, and from hematite by streak. **Occurrence** Ilmenite is an accessory mineral in igneous rocks such as gabbro and diorite. It occurs occasionally in quartz veins and pegmatites with hematite and chalcopyrite, and in some gneisses. Its resistance to weathering leads to its concentration in alluvial sands, together with magnetite, monazite, and rutile. The name is taken from Lake Ilmen, Russia.

Ilmenite

Chrysoberyl $BeAl_2O_4$

Crystal system Orthorhombic. **Habit** Crystals generally tabular. **Twinning** Common, often repeated, to give pseudo-hexagonal crystals. **SG** 3.5–3.8 **Hardness** $8\frac{1}{2}$ **Cleavage** Prismatic, poor. **Fracture** Uneven to conchoidal. **Color and transparency** Various shades of green and yellow: transparent to translucent. Chrysoberyl is used as a gemstone when transparent; the variety alexandrite is emerald green but red by artificial light.

Chrysoberyl: twinned crystal

Ilmenite

Chrysoberyl

2 ins

46

Streak White. **Luster** Vitreous. **Distinguishing features** Color and hardness; crystals tabular, in contrast to the prismatic habit of beryl. May be confused with olivine. **Occurrence** Chrysoberyl occurs in granitic rocks and pegmatites; also in mica schists. It is frequently found in alluvial sands and gravels. The name is taken from the Greek and alludes to the golden yellow color. Alexandrite was named for Tsar Alexander II of Russia.

Corundum

Corundum

2 ins

Corundum
(star sapphire)

Corundum Al_2O_3

Crystal system Trigonal. **Habit** Crystals usually rough, barrel-shaped prismatic forms or tapering, spindle-shaped forms; also flat, tabular. Emery is a mixture of the massive black granular form with magnetite and spinel. **Twinning** Common, often repeated, giving rise to striations on basal pinacoid. **SG** 3.9–4.1 **Hardness** 9 **Cleavage** None; basal parting. **Fracture** Uneven to conchoidal. **Color and transparency** Two main varieties; blue (sapphire), red (ruby). Also yellow, brown, green; crystals sometimes show variation in color. Star sapphire and star ruby are opalescent and show a six-rayed star when polished and viewed in one particular direction. Transparent to translucent. **Streak** White. **Luster** Adamantine to vitreous. **Distinguishing features** Great hardness, specific gravity, crystal form. **Occurrence** Corundum occurs in certain nepheline syenites and nepheline syenite pegmatites. It occurs also in metamorphic rocks such as marble, gneiss, and schist. Large crystals occur in some pegmatites, and emery deposits occur in some regionally metamorphosed rocks. The hardness and durability of corundum lead to its occurrence in alluvial sands and gravels. Gem-quality ruby occurs in Burma (Myanmar) and Sri Lanka. Sapphire also occurs there and in India, Australia, and elsewhere. Corundum is also used as an abrasive either as fragments produced by grinding massive corundum, or as the impure form, emery.

Corundum

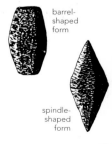

barrel-shaped form

spindle-shaped form

47

Pyrochlore-microlite series $(Ca,Na)_2(Nb,Ta)_2O_6(O,OH,F)$

Crystal system Cubic. **Habit** Crystals commonly octahedral; also as small, irregular grains. **SG** 4.2–6.4 Increasing with tantalum content. **Hardness** $5-5\frac{1}{2}$ **Cleavage** Octahedral, distinct. **Fracture** Conchoidal. **Color and transparency** Pyrochlore is brown to black; microlite yellow to brown, sometimes red: subtranslucent to opaque. **Streak** Light brown. **Luster** Vitreous to resinous; sometimes greasy. **Distinguishing features** Crystal form. Pyrochlore is the name given to the niobium-rich members of the series and microlite to those in which tantalum is dominant. Many other elements can substitute for sodium and calcium in the structure, including uranium, thorium and the rare earth elements. Some specimens are therefore radioactive. **Occurrence** Usually in pegmatites in, or near to, alkaline rocks and associated with zircon and apatite. Pyrochlore occurs characteristically in pipe-like igneous intrusions composed essentially of calcium and magnesium carbonates and called carbonatite. Microlite is usually associated with granite pegmatites. The name pyrochlore is derived from the Greek in allusion to the green color it acquires on heating, and microlite is so named because of the small size of the first described crystals.

Pyrochlore

2 ins

Microlite

Braunite $Mn^{2+}Mn^{3+}_6SiO_{12}$

Crystal system Tetragonal. **Habit** Pyramidal crystals; also massive, granular. Braunite departs only slightly from cubic symmetry and so the crystals appear octahedral. **SG** 4.7–4.8 **Hardness** $6-6\frac{1}{2}$ **Cleavage** Pyramidal, perfect. **Fracture** Uneven. **Color and transparency** Brownish black to steel-gray: opaque. **Streak** Brownish black to steel-gray. **Luster** Submetallic. **Distinguishing features** Color, crystal form. Soluble in hydrochloric acid, leaving a residue of silica. **Occurrence** Occurs in hydrothermal veins with other manganese oxides, and also as a secondary mineral. It sometimes forms as a result of metamorphism of manganese-bearing sediments. It is named after K. Braun von Gotha.

Psilomelane

Braunite

Wad

2 ins

Psilomelane No fixed composition: a mixture of manganese oxides; **Romanèchite** $(Ba,H_2O)_2(Mn^{4+},Mn^{3+})_5 O_{10}$
Crystal system (romanèchite) Monoclinic. **Habit** Massive, botryoidal, stalactitic. **SG** 3.3–4.7 **Hardness** 5–7 **Color and transparency** Black to dark gray: opaque. **Streak** Brownish black to black. **Luster** Submetallic. **Distinguishing features** Psilomelane refers to the group of massive, black manganese oxides. The original psilomelane is now called romanèchite. Greater hardness distinguishes psilomelane from other manganese minerals; botryoidal habit. **Occurrence** Secondary manganese minerals, precipitated at atmospheric temperatures together with pyrolusite and limonite, in sediments or in quartz veins.

Psilomelane: botryoidal stalactitic form

Wad No fixed composition
Crystal system Amorphous. **Habit** Shapeless, stalactitic, or reniform masses, often earthy. **SG** 2.8–4.4 **Hardness** Usually soft. **Color and transparency** Dull black to brownish black. **Streak** Black. **Luster** Earthy. **Distinguishing features** Black color; often loosely compacted and so feels light and soils the fingers. **Occurrence** Wad is not a single mineral but a mixture of several hydrous manganese oxides which occur in the oxidized zone of ore deposits, in lake or bog deposits, and shallow water marine sediments.

Cassiterite (Tinstone) SnO_2

Crystal system Tetragonal. **Habit** Crystals pyramidal or short prismatic; also massive, granular. Reniform shapes with a fibrous structure are called wood tin. **Twinning** Common. **SG** 6.8–7.1 **Hardness** 6–7 **Cleavage** Prismatic, imperfect. **Fracture** Uneven. **Color and transparency** Usually reddish brown to nearly black; sometimes yellowish: opaque to translucent. **Streak** White or grayish. **Luster** Adamantine, splendent to submetallic. **Distinguishing features** High specific gravity, adamantine luster, light streak, crystal form. **Occurrence** Cassiterite is one of the few tin minerals and is the principal ore of the metal. It occurs typically in high temperature hydrothermal veins and pegmatites located within, or close to, granite masses. Associated minerals are wolframite, arsenopyrite, bismuthinite, topaz, quartz, tourmaline, and mica. Rounded pebbles of cassiterite occur as stream tin in alluvial deposits.

Cassiterite

Cassiterite

2 ins

Pyrolusite MnO_2

Crystal system Tetragonal. **Habit** Usually massive, often as reniform coatings or dendritic, plant-like shapes on joints or bedding planes in sedimentary rocks; also as divergent fibers or columns. Very rarely as crystals (polianite). **SG** 4.5–7.9 **Hardness** 1–2 (massive), 6–6½ (crystals) **Cleavage** Perfect prismatic. **Fracture** Uneven, splintery. **Color and transparency** Black to bluish steel-gray: opaque. **Streak** Black. **Luster** Metallic. **Distinguishing features** Color, softness when massive. **Occurrence** Pyrolusite is a common manganese mineral and forms under oxidizing conditions. It is often of secondary origin, being found in the oxidized zone of ore deposits containing manganese, and in bog or shallow marine sediments. It occurs also in quartz veins, and as nodules on the bottom of the sea. The name is derived from two Greek words meaning "fire" and "to wash", because it was at one time used to remove from glass the color due to the presence of iron oxide.

Pyrolusite: dendritic form

Rutile TiO_2

Crystal system Tetragonal. **Habit** Crystals prismatic and terminated by bipyramids. Prism faces often striated. Also massive. **Twinning** Common on bipyramid giving geniculate (knee-shaped) twins, or complex cyclic twins made up of six or eight individuals. **SG** 4.2–4.4 **Hardness** $6-6\frac{1}{2}$ **Cleavage** Prismatic, distinct. **Fracture** Uneven. **Color and transparency** Usually reddish brown; can be yellowish red or black. Transparent when thin, usually subtranslucent, occasionally nearly opaque. **Streak** Pale brown. **Luster** Adamantine; submetallic when dark colored. **Distinguishing features** Red color, adamantine luster, crystal form. **Occurrence** Rutile occurs as an accessory mineral in a variety of igneous rocks, and in schists, gneisses, metamorphosed limestones, and quartzites. It often occurs as acicular crystals in quartz (rutilated quartz). It is also a secondary mineral produced by the breakdown of titanium-bearing minerals such as titanite and some micas. Rutile is also concentrated in alluvial deposits and beach sands.

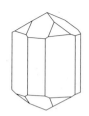

Rutile

Rutile: geniculate twin

Pyrolusite

2 ins

Pyrolusite

Rutile

Anatase

Anatase (Octahedrite) TiO$_2$
Crystal system Tetragonal. **Habit** Crystals commonly bipyrami-dal; also tabular. **SG** 3.8–4.0 **Hardness** 5$\frac{1}{2}$–6 **Cleavage** Basal and bipyramidal, perfect. **Fracture** Subconchoidal. **Color and transparency** Varies from shades of yellow and brown to blue and black: transparent to nearly opaque. **Streak** White. **Luster** Adamantine, becoming metallic when black and opaque. **Distinguishing features** Crystal form, frequent association with rutile and brookite. **Occurrence** Anatase is an accessory mineral in igneous and metamorphic rocks, having been derived from other titanium minerals and deposited by hydrothermal solu-tions. It occurs also in granite pegmatites.

Brookite

2 ins

Anatase: bipyramidal habit

Columbite: tantalite

Brookite TiO$_2$
Crystal system Orthorhombic. **Habit** Crystals vary in habit but often tabular or platy. **SG** 3.9–4.2 **Hardness** 5$\frac{1}{2}$–6 **Cleavage** Prismatic, poor. **Fracture** Subconchoidal to uneven. **Color and transparency** Reddish brown to brownish black: translucent. **Streak** White. **Luster** Metallic-adamantine. **Distinguishing fea-tures** Crystal form, luster, frequent association with rutile and anatase. **Occurrence** As an accessory mineral in igneous and metamorphic rocks, and in hydrothermal veins. Rutile, anatase and brookite are polymorphs of TiO$_2$. Brookite is named after H. J. Brooke (1771–1857), a British mineralogist.

Columbite-tantalite series (Fe,Mn,Mg)(Nb,Ta)$_2$O$_6$
Crystal system Orthorhombic. **Habit** Crystals tabular or short prismatic. **Twinning** Common. **SG** 5.0–8.0 (increases with tantalum content). **Hardness** 6–6$\frac{1}{2}$ **Cleavage** Pinacoidal. **Frac-ture** Uneven. **Color and transparency** Iron-black to brownish black: subtranslucent to opaque. **Streak** Dark red to black. **Luster** Submetallic to subresinous. **Distinguishing features** High specific gravity, black color, crystal form. As with the pyrochlore-microlite series there is continuous substitution of tantalum for niobium. When niobium dominates over tantalum, the general name columbite is applied (covering the species ferrocolumbite, magnocolumbite and manganocolumbite), and tantalite (cover-ing ferrotantalite and manganotantalite) when tantalum is present in greater amount than niobium. **Occurrence** Usually in granite pegmatites in association with quartz, feldspar, tour-maline, beryl, spodumene, petalite, cassiterite, and wolframite. Columbite and tantalite are sources of niobium and tantalum.

Columbite-
tantalite

Uraninite

2 ins

Uraninite (Pitchblende) UO_2
Crystal system Cubic. **Habit** Crystals rare: usually massive,
botryoidal (pitchblende). **SG** 6.5–8.5 (massive), 8–10 (crystals)
Hardness 5–6 **Fracture** Conchoidal to uneven. **Color and
transparency** Brownish black to black: opaque. **Streak** Brownish
black or gray. **Luster** Submetallic to greasy or pitch-like; dull.
Distinguishing features High specific gravity, characteristic
greasy or pitch-like luster, black color, radioactivity. The name
uraninite is used for crystallized varieties of UO_2; massive varieties
are called pitchblende. **Occurrence** Crystallized uraninite occurs
in pegmatites associated with granite or syenitic rocks and
associated with monazite, zircon, and tourmaline. Pitchblende is
usually found as massive crusts in high- or moderate-temperature
hydrothermal veins associated with cassiterite, pyrite, chalcopyrite, arsenopyrite, and galena. It occurs also as a detrital mineral
in alluvial deposits. It is an important ore of uranium, and was the
source of radium, discovered by the Curies.

Brucite $Mg(OH)_2$
Crystal system Trigonal. **Habit** Crystals broad tabular; also fibrous (nemalite) and massive, foliated. **SG** 2.4 **Hardness** $2\frac{1}{2}$ **Cleavage** Basal, perfect. **Color and transparency** White, shading to pale gray, blue or green: transparent to translucent. **Streak** White. **Luster** Pearly parallel to cleavage, elsewhere waxy to vitreous. **Distinguishing features** Cleavage, softness, foliated habit. Distinguished from talc by greater hardness, and from gypsum by form. Readily soluble in hydrochloric acid, much more so than gypsum. **Occurrence** Brucite occurs in metamorphosed dolomitic limestones and also in hydrothermal veins together with calcite and talc, and in serpentinite. It is named in honor of A. Bruce (1777–1818), an American mineralogist.

Gibbsite (Hydrargillite) $Al(OH)_3$
Crystal system Monoclinic. **Habit** Crystals tabular; also as stalactitic or encrusting forms, spheroidal concretions or foliated and earthy aggregates. **Twinning** Common. **SG** 2.3–2.4 **Hardness** $2\frac{1}{2}-3\frac{1}{2}$ **Cleavage** Basal, perfect. **Color and transparency** White or near white, sometimes pink or red: transparent to translucent. **Streak** White. **Luster** Pearly parallel to cleavage; other faces vitreous. **Distinguishing features** Strong clay odor when breathed on. **Occurrence** Gibbsite occurs as crystals in low-temperature hydrothermal veins, and with böhmite and diaspore as a constituent mineral of bauxite. It is a secondary mineral resulting from the decomposition of aluminum silicates. It is named after Col. G. Gibbs (1776–1833), an American mineral collector.

Brucite

Gibbsite

Gibbsite

Gibbsite

2 ins

Böhmite AlO(OH)

Crystal system Orthorhombic. **Habit** Crystals microscopic; usually as scattered grains or pisolitic aggregates. **SG** 3.0–3.1 **Hardness** $3\frac{1}{2}$–4 **Cleavage** One very good cleavage. **Color** White. **Occurrence** Böhmite, with gibbsite, diaspore and kaolinite, is an important constituent mineral of bauxite, but because of its occurrence as tiny crystals or grains, it cannot be recognized by the naked eye. It is named after J. G. Böhm, a 19th-century German chemist.

Diaspore AlO(OH)

Crystal system Orthorhombic. **Habit** Crystals platy or tabular; also massive, foliated, sometimes acicular. **SG** 3.2–3.5 **Hardness** $6\frac{1}{2}$–7 **Cleavage** One perfect cleavage. **Color and transparency** Variable, from colorless, through white and gray, to brown or pink: translucent. **Luster** Pearly on cleavage faces, vitreous elsewhere. **Distinguishing features** Cleavage, hardness, platy habit. **Occurrence** Diaspore is an important constituent of bauxite together with böhmite and gibbsite. It occurs in close association with corundum in emery deposits, and in chlorite schist.

Lepidocrocite Fe^{3+}O(OH)

Crystal system Orthorhombic. **Habit** Scaly, fibrous or massive aggregates. **SG** 4.1 **Hardness** 5 **Cleavage** One perfect cleavage. **Color** Red to reddish brown. **Streak** Orange. **Distinguishing features** Color. **Occurrence** Lepidocrocite and goethite are similar in chemistry and occurrence and it is difficult to distinguish them.

Lepidocrocite

2 ins

Böhmite

Diaspore

Goethite Fe^{3+}O(OH)

Crystal system Orthorhombic. **Habit** Crystals rare but platy, bladed or prismatic; usually massive as mammillary, botryoidal or stalactitic masses with a fibrous radiating structure. **SG** 3.3–4.3 **Hardness** 5–5$\frac{1}{2}$ **Cleavage** One perfect cleavage. **Fracture** Uneven. **Color and transparency** Usually very dark brown; earthy forms ochrous yellow-brown: subtranslucent; transparent in thin fragments. **Streak** Brownish yellow. **Luster** Adamantine (crystals); massive goethite is often silky by reason of its fibrous structure; sometimes dull. **Distinguishing features** Color, streak. The presence of cleavage, radial growth, and other indications of crystallinity distinguish goethite from limonite. **Occurrence** Goethite is produced by the oxidation of iron-bearing minerals such as pyrite and magnetite. It is of widespread occurrence, though at one time it was thought to be a rare mineral. In fact much of the ochrous brown ferric oxide described as "limonite" is composed in large part of crystalline goethite. Limonite has a yellowish brown streak and a vitreous luster which, together with

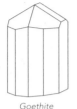

Goethite

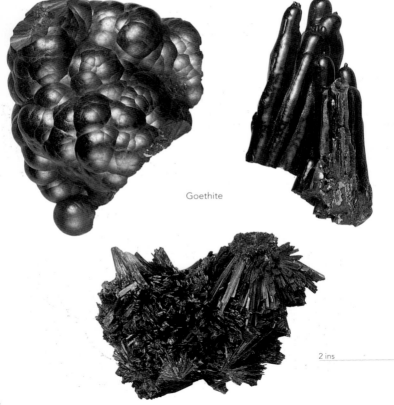

Goethite

2 ins

the lack of cleavage, serves to distinguish it from crystalline goethite. Goethite occurs as a secondary mineral in the oxidized zone (iron hat) of veins containing iron-bearing minerals. It replaces other minerals, and goethite pseudomorphs after pyrite are common. Oolitic "limonitic" iron ores of sedimentary origin occur in eastern France (the minette ores), and goethite is precipitated from marine or fresh water in bogs or lagoons to form bog iron ore. The mineral is named after the German poet J. W. von Goethe (1749–1832), who was also a mineral collector.

Limonite

Manganite Mn^{3+}O(OH)
Crystal system Monoclinic, pseudo-orthorhombic. **Habit** Crystals prismatic, often striated and frequently grouped in bundles or as radiating aggregates. **Twinning** Penetration twins on prism. **SG** 4.2–4.4 **Hardness** 4 **Cleavage** Pinacoidal, perfect; prismatic, less so. **Fracture** Uneven. **Color and transparency** Dark steel-gray to black: opaque. **Streak** Reddish brown to black. **Luster** Submetallic. **Distinguishing features** Color, prismatic habit, brown streak, soluble in concentrated hydrochloric acid. **Alteration** To pyrolusite and other manganese oxides. **Occurrence** Manganite occurs in association with such minerals as pyrolusite, barite and goethite, in deposits precipitated from waters under oxidizing conditions. It occurs also in low-temperature hydrothermal veins, associated with granitic rocks. Manganite has been used as an ore of manganese.

Manganite

Bauxite
Habit Massive, oolitic, pisolitic, earthy; or as concretionary masses. **Color** Ocher-yellow, brown, red, also gray. **Distinguishing features** Color; earthy, oolitic to pisolitic structure. **Occurrence** Bauxite is not a single mineral but a mixture of several minerals, mainly diaspore, gibbsite, böhmite and iron oxides (see pages 54 and 55). It is of secondary origin, forming under tropical conditions by the prolonged weathering and leaching of rocks containing aluminum silicates. The leaching by tropical rains removes the silica, leaving aluminous hydroxides.

Bauxite: pisolitic structure

2 ins

Manganite

Halite: cube

Halite: hopper crystal

Halite (Rock salt) NaCl
Crystal system Cubic. **Habit** Usually as cubes, often with concave faces (hopper crystals); also massive, granular, compact (rock salt). **SG** 2.1–2.2 (2.16 when pure) **Hardness** $2\frac{1}{2}$ **Cleavage** Cubic, perfect. **Fracture** Conchoidal. **Color and transparency** Colorless or white, but also shades of yellow, red, and sometimes blue: transparent to translucent. **Streak** White. **Luster** Vitreous. **Distinguishing features** Ready solubility in water, perfect cubic cleavage, salty taste. Taste, for obvious reasons, is rarely used as a test in mineral identification. It is, however, a useful confirmatory test for halite. **Occurrence** Halite is widely distributed in stratified evaporite deposits formed when enclosed saline waters evaporate, for example around present-day playa lakes. Beds of halite, associated with other water-soluble minerals, such as sylvite, gypsum, and anhydrite, occur in sedimentary basins of various ages, having formed by the evaporation of land-locked seas in the geological past. Not infrequently, plug-like masses (salt domes) rise from the salt layer and intrude and arch up the overlying sediments, sometimes forming oil traps.

Halite

2 ins

58

Sylvite

Cryolite

Sylvite KCl

Crystal system Cubic. **Habit** Crystals usually cubic but often as combination of cube and octahedron; also massive, compact. **SG** 2.0 **Hardness** 2 **Cleavage** Cubic, perfect. **Fracture** Uneven. **Color and transparency** Colorless or white, sometimes also shades of blue, yellow or red: transparent to translucent. **Streak** White. **Luster** Vitreous. **Distinguishing features** Sylvite is similar to halite, with which it is commonly associated. It is best distinguished from halite by its bitter taste. **Occurrence** Sylvite, like halite, occurs in bedded evaporite deposits, but is present in lesser amounts because of its greater solubility in water.

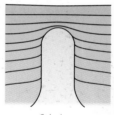

Salt dome

Cryolite Na$_3$AlF$_6$

Crystal system Monoclinic. **Habit** Crystals rare and pseudo-cubic in appearance, combinations of prisms and pinacoids resembling cube and octahedron; also massive, granular. **Twinning** Common; complex. **SG** 3.0 **Hardness** $2\frac{1}{2}$ **Cleavage** None; basal and prismatic parting. **Fracture** Uneven. **Color and transparency** Colorless to white, sometimes brownish to reddish: transparent to translucent. **Streak** White. **Luster** Vitreous to greasy. **Distinguishing features** Pseudo-cubic cleavage, greasy luster, fuses readily when heated. Cryolite has a low refractive index (about 1.34) so that it becomes almost invisible when its powder is placed in water. **Occurrence** Cryolite is a rare mineral which is used as a flux in the production of aluminum by the electrolytic process. The only notable locality is in Greenland, where it occurs in pegmatites associated with granite, and in company with siderite, quartz, galena, sphalerite, chalcopyrite, fluorite, cassiterite, and many other minerals. The name is derived from Greek words meaning "ice-stone" and refers to its ice-like appearance.

Cryolite

2 ins

Carnallite

2 ins

Chlorargyrite

Carnallite KMgCl$_3$.6H$_2$O
Crystal system Orthorhombic. **Habit** Crystals rare and pseudo-hexagonal; usually massive, granular. **SG** 1.6 **Hardness** 1–2 **Cleavage** None. **Fracture** Conchoidal. **Color and transparency** White, sometimes reddish or yellowish: transparent to translucent. **Luster** Greasy. **Distinguishing features** Lack of cleavage, conchoidal fracture, deliquescence (absorbs moisture and becomes wet), bitter taste, fuses readily when heated. **Occurrence** Carnallite occurs in evaporite deposits together with minerals such as halite and sylvite. Specimens of carnallite should be kept in sealed bottles because of their deliquescent nature. The mineral is named after R. von Carnall, a 19th-century German mining engineer.

Chlorargyrite (Horn silver, Cerargyrite) AgCl
Crystal system Cubic. **Habit** Crystals rare but usually cubes; commonly massive, resembling wax or horn. **Twinning** On octahedron. **SG** 5.5–5.6 **Hardness** 1$\frac{1}{2}$–2$\frac{1}{2}$ **Cleavage** None. **Fracture** Subconchoidal. **Color and transparency** Colorless when pure; usually pearl-gray becoming violet-brown on exposure to light: translucent. **Luster** Resinous to adamantine. **Distinguishing features** Hornlike appearance, can be cut with a knife, melts readily when heated, soluble in ammonium hydroxide. **Occurrence** Chlorargyrite occurs as a secondary mineral in the oxidized zone of silver deposits, together with native silver and cerussite.

Atacamite Cu$_2$Cl(OH)$_3$
Crystal system Orthorhombic. **Habit** Crystals slender, prismatic and striated; or tabular; also massive, fibrous, granular. **Twinning** Complex. **SG** 3.8 **Hardness** 3–3½ **Cleavage** Pinacoidal, perfect. **Fracture** Conchoidal. **Color and transparency** Bright green to dark green: transparent to translucent. **Streak** Apple-green. **Luster** Adamantine to vitreous. **Distinguishing features** Color; distinguished from malachite by lack of effervescence in hydrochloric acid, though readily soluble. **Occurrence** Atacamite is always of secondary origin, forming in the oxidized zone of copper deposits, and commonly associated with cuprite and malachite. Named after Atacama, Chile.

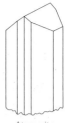

Atacamite

Diaboleite Pb$_2$CuCl$_2$(OH)$_4$
Crystal system Tetragonal. **Habit** Crystals tabular; also massive, granular. **SG** 5.5 **Hardness** 2½ **Cleavage** Basal. **Color and transparency** Bright blue: translucent to nearly opaque. **Distinguishing features** Color. **Occurrence** Diaboleite is a rare but colorful mineral that occurs in the oxidized parts of some copper-lead ore deposits.

Diaboleite

Atacamite

Atacamite

2 ins

Diaboleite

Fluorite

2 ins

Fluorite: modified cube

Fluorite: modified cube

Fluorite (Fluorspar) CaF_2
Crystal system Cubic. **Habit** Crystals commonly cubic, less frequently octahedral or rhombdodecahedral. Combinations of cube with octahedron or rhombic dodecahedron often have cube faces smooth and others dull, or rough, being formed of tiny cube faces in parallel arrangement. **Twinning** Interpenetration twins common. **SG** 3.2 **Hardness** 4 **Cleavage** Octahedral, perfect. **Fracture** Subconchoidal. **Color and transparency** Color varies greatly; it is often yellow, green, blue, purple; more rarely colorless, pink, red, and black: transparent to translucent. Single crystals may vary in color and, like some fluorite masses, are often color-banded. **Streak** White. **Luster** Vitreous. **Distinguishing features** Cubic crystal form, octahedral cleavage, harder than calcite, lack of effervescence with hydrochloric acid. Dissolves in sulfuric acid giving off fumes of hydrogen fluoride which etch glass. Fluorescent. **Occurrence** Fluorite is a widely distributed mineral. It occurs in mineral veins, either alone or as

Fluorite

2 ins

a gangue mineral with metallic ores, and in association with quartz, barite, calcite, celestine, dolomite, galena, cassiterite, sphalerite, topaz, and many other minerals. Although fluorite is too soft and too readily cleavable to be used as a precious stone, the color variation, particularly in the color-banded variety known as Blue John, has made it prized as a semiprecious ornamental stone from which vases and ornaments have been fashioned since ancient times. Fluorite has given its name to the phenomenon of fluorescence but it shows this effect only weakly. Many other minerals are more spectacular in this respect. Fluorite is worked mainly for use as a flux in the smelting of iron and in the chemical industry; smaller amounts are used as decorative stones and in the manufacture of specialized optical equipment. The name comes from the Latin word *fluere* meaning "to flow" in reference to its low melting point and use as a flux in the smelting of metals.

Fluorite:
interpenetration twin

Calcite

2 ins

Calcite CaCO₃

Crystal system Trigonal. **Habit** Crystals common and very varied in habit, more so than any other mineral. The commonest habits are tabular; prismatic; acute or obtuse rhombohedral, and scalenohedral (dog tooth spar). Calcite also occurs as parallel or fibrous aggregates, or as granular, stalactitic or massive aggregates. **Twinning** Common; there are two main laws, in the first the twin plane is the basal pinacoid, and in the second the twin plane is a rhombohedral face. Lamellar twinning may also be produced by pressure. **SG** 2.7 (when pure) **Hardness** 3 **Cleavage** Rhombohedral, perfect. Although the habit of calcite is so variable, it always cleaves into rhombohedral cleavage fragments. The phenomenon known as double refraction is well shown by calcite, in that if a clear cleavage rhomb of calcite is placed over a dark spot on paper, two images are seen. If the calcite rhomb is rotated, one image remains stationary, and the other rotates with the rhomb. **Fracture** Conchoidal but rarely seen owing to perfection of cleavage. **Color and transparency** Usually colorless (Iceland spar) or white; also shades of gray, yellow, green, red, purple, blue, and even brown or black: transparent to translucent; some deeply colored forms nearly opaque. **Streak** White. **Distinguishing features** Perfect rhombohedral cleavage, hardness, dissolves readily with effervescence in cold dilute hydrochloric

Calcite: prism and flat rhombohedron

Calcite: rhombohedron

Calcite: scalenohedron

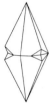

Calcite: scalenohedron twinned on basal pinacoid

Calcite: scalenohedron twinned on rhombohedron

Calcite: scalenohedron twinned on rhombohedron

acid. **Occurrence** Calcite is a common and widely distributed mineral. It is a rock-forming mineral that is a major constituent in calcareous sedimentary rocks (limestone) and metamorphic rocks (marble). It may be precipitated directly from sea water, and it forms the shells of many living organisms which, on death, accumulate to form limestone. Metamorphosed limestone, when pure, forms white granular marble; the presence of other minerals results in colored, figured marble. It occurs also as a primary mineral in carbonatites, which are calcareous igneous rocks that form intrusive plugs. It is of common occurrence in veins, either as the main constituent or as a gangue mineral accompanying metallic ores. Secondary calcite sometimes replaces primary minerals such as pyroxenes or feldspars in igneous rocks. In areas of hot springs it is deposited as travertine or tufa, and stalactites and stalagmites of calcite are common in caves in limestone areas. Calcite, in the form of limestone, is quarried on a large scale for use in the making of cement, as a flux in the smelting of metallic ores, as a fertilizer and as a building stone.

Calcite: combination of prism, scalenohedron, and rhombohedron

Siderite

Siderite

2 ins

Magnesite

Magnesite MgCO₃

Crystal system Trigonal. **Habit** Crystals uncommon but rhombohedral or prismatic; usually massive, granular, compact or fibrous. **SG** 3.0–3.2 (increasing with iron content) **Hardness** 3½–4½ **Cleavage** Rhombohedral, perfect. **Fracture** Conchoidal. **Color and transparency** White or colorless when pure; often in grayish, yellowish or brownish shades when iron is present: transparent to translucent. **Streak** White. **Luster** Vitreous. **Distinguishing features** Similar to calcite; hardly affected by cold dilute hydrochloric acid but dissolves with effervescence when warmed. **Occurrence** Magnesite is much less common than calcite and does not usually form sedimentary rocks. It occurs as replacement deposits formed by the action of carbonate-bearing waters on rocks containing magnesium minerals, or by the action of magnesium-rich solutions on calcite-bearing rocks. Magnesite occurs also as veins in magnesium-rich metamorphic rocks such as talc schists and serpentinites. Extensive deposits of replacement origin are worked for use in making cement and refractory bricks.

Siderite (Chalybite) $FeCO_3$

Crystal system Trigonal. **Habit** Crystals rhombohedral, usually with curved faces which are composite, being an aggregate of small individuals. Also massive, granular, fibrous, compact, botryoidal or earthy. **Twinning** On rhombohedron; often lamellar. **SG** 3.8–4.0 (decreasing with magnesium content) **Hardness** $3\frac{1}{2}$–$4\frac{1}{2}$ **Cleavage** Rhombohedral, perfect. **Fracture** Uneven. **Color and transparency** Gray to gray-brown and yellowish brown: transparent to translucent. **Streak** White. **Luster** Vitreous. **Distinguishing features** Rhombohedral form and cleavage; its brown color and higher specific gravity distinguish it from calcite and dolomite. Dissolves slowly in cold dilute hydrochloric acid but dissolves with effervescence when warmed. **Occurrence** Massive siderite is widespread in sedimentary rocks, particularly in clays and shales where it forms clay ironstones which are usually of concretionary origin *(see pages 208–209)*. It also occurs as a gangue mineral in hydrothermal veins accompanying metallic ore minerals such as pyrite, chalcopyrite and galena; and it also forms where limestone has been replaced by the action of iron-bearing solutions.

Rhodochrosite $MnCO_3$

Crystal system Trigonal. **Habit** Crystals rare but rhombohedral and with curved faces; usually massive, compact, cleavable or granular. **SG** 3.4–3.7 **Hardness** $3\frac{1}{2}$–$4\frac{1}{2}$ **Cleavage** Rhombohedral, perfect. **Fracture** Uneven. **Color and transparency** Rose-pink, sometimes light gray to brown: translucent. **Streak** White. **Luster** Vitreous. **Distinguishing features** Color, rhombohedral cleavage; dissolves with effervescence in hot dilute hydrochloric acid. It is distinguished from rhodonite by its inferior hardness, and often develops a brown or black crust on exposure to air. **Occurrence** Rhodochrosite occurs in hydrothermal mineral veins containing ores of silver, lead, and copper. It has been also noted in metamorphic and metasomatic rocks of sedimentary origin, and in sedimentary deposits of manganese oxide, where it is of secondary origin.

2 ins

Rhodochrosite

2 ins

Smithsonite

Smithsonite

Dolomite: showing curved composite faces

Smithsonite (Calamine) $ZnCO_3$

Crystal system Trigonal. **Habit** Crystals rare but rhombohedral with rough, curved faces; usually as botryoidal, reniform, stalactitic or encrusting masses; also massive. **SG** 4.3–4.5 **Hardness** 4–$4\frac{1}{2}$ **Cleavage** Rhombohedral, perfect. **Fracture** Uneven. **Color and transparency** Shades of gray, brown or grayish white but green, brown and yellow varieties also occur: translucent. **Streak** White. **Luster** Vitreous. **Distinguishing features** Rhombohedral cleavage, high density for a carbonate, soluble in dilute hydrochloric acid with effervescence. **Occurrence** The main occurrence of smithsonite is in the oxidized zone of ore deposits carrying zinc minerals. It is commonly associated with sphalerite, hemimorphite, galena and calcite. It is recorded also in some hydrothermal veins accompanying sphalerite, and as a replacement of limestone. The translucent green variety of smithsonite is used as an ornamental stone. It is named after J. L. M. Smithson (1754–1829), a British mineralogist, and founder of the Smithsonian Institution in Washington, DC, USA.

Dolomite $CaMg(CO_3)_2$

Crystal system Trigonal. **Habit** Crystals usually rhombohedral with curved composite faces; also occurs in massive, granular aggregates and as a rock-forming mineral in dolomitic limestones. **Twinning** Common. **SG** 2.8–2.9 **Hardness** $3\frac{1}{2}$–4 **Cleavage** Rhombohedral, perfect. **Fracture** Subconchoidal. **Color and transparency** Usually white, though sometimes colorless; also yel-

2 ins

Dolomite

Ankerite

Dolomite

lowish to brown, occasionally pink: transparent to translucent. **Streak** White. **Luster** Vitreous to pearly. **Distinguishing features** Dolomite is similar to calcite but dissolves only slowly in cold dilute hydrochloric acid, but effervesces readily when warmed. **Occurrence** Dolomite occurs widely as a rock-forming mineral. It is usually of secondary occurrence, having formed by the action of magnesium-bearing solutions on limestone. It also occurs as a gangue mineral in hydrothermal veins, particularly those containing galena and sphalerite. Dolomitic limestones are used as building stones, and the mineral is used in the manufacture of refractory bricks for furnace linings. Dolomite is named after D. Dolomieu (1750–1801), a French mineralogist.

Ankerite $Ca(Mg,Fe,Mn)(CO_3)_2$

Crystal system Trigonal. **Habit** Crystals rhombohedral; also massive, granular. **SG** 2.9–3.2 **Hardness** $3\frac{1}{2}$–4 **Cleavage** Rhombohedral. **Color and transparency** White, yellow, yellowish brown, sometimes gray; becomes dark brown on weathering: translucent. **Streak** White. **Luster** Vitreous. **Distinguishing features** Brown color. Ferrous iron substitutes for magnesium in dolomite, and there is a series from dolomite to ankerite, with the brown color usually becoming more pronounced with increasing iron content. **Occurrence** Ankerite occurs in similar ways to dolomite; it is often a gangue mineral accompanying iron ores, and it frequently fills joints in coal seams. Ankerite is named after M. J. Anker (1772–1843), an Austrian mineralogist.

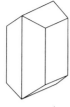

Aragonite

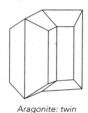

Aragonite: twin

Aragonite: repeated twin

Aragonite CaCO₃

Crystal system Orthorhombic. **Habit** Untwinned crystals, which are rare, are often acicular, though sometimes tabular. Twins stout, prismatic, with marked pseudo-hexagonal symmetry. Also as fibrous, stalactitic and encrusting masses. **Twinning** Very common. Repeated twinning produces pseudo-hexagonal forms. **SG** 2.9 **Hardness** $3\frac{1}{2}$–4 **Cleavage** Pinacoidal, imperfect. **Fracture** Subconchoidal. **Color and transparency** Colorless, gray, white; also yellowish: transparent to translucent. **Streak** White. **Luster** Vitreous. **Distinguishing features** Soluble with effervescence in cold dilute hydrochloric acid. Aragonite is a polymorph of $CaCO_3$ and is distinguished from calcite by its form, lack of rhombohedral cleavage, and greater specific gravity. **Occurrence** Aragonite is not as widespread as its polymorph, calcite. It occurs as a deposit from hot springs and in association with beds of gypsum. It has been noted in veins and cavities with calcite and dolomite, and in the oxidized zone of ore deposits together with secondary minerals such as malachite and smithsonite. The shells of certain mollusks are made of aragonite and many fossil shells now composed of calcite were probably formed originally of aragonite. It also occurs in some glaucophane schists in association with jadeite and glaucophane. The name comes from the province of Aragon, in Spain, where it was first noted.

Witherite BaCO₃

Crystal system Orthorhombic. **Habit** Crystals invariably twinned with pseudo-hexagonal form; also massive, granular, columnar, botryoidal. **Twinning** Ubiquitous. **SG** 4.3 **Hardness** $3-3\frac{1}{2}$ **Cleavage** Pinacoidal, distinct. **Fracture** Uneven. **Color and transparency** White; sometimes gray or pale yellow to brown: transparent to translucent. **Streak** White. **Luster** Vitreous. **Distinguishing features** High specific gravity, soluble in dilute hydrochloric acid with effervescence. Distinguished from strontianite by the flame test; witherite colors the flame green. **Occurrence** Witherite is not of wide occurrence. It sometimes accompanies galena in hydrothermal veins, together with anglesite and barite. It is named after W. Withering (1741–99), the British mineralogist who first recognized and analysed the mineral.

Strontianite SrCO₃

Crystal system Orthorhombic. **Habit** Crystals prismatic or acicular; also massive, fibrous, columnar, granular. **Twinning** Common. **SG** 3.7 **Hardness** $3\frac{1}{2}$–4 **Cleavage** Prismatic, good. **Fracture** Uneven. **Color and transparency** White, pale green, gray, pale yellow: transparent to translucent. **Streak** White. **Luster** Vitreous. **Distinguishing features** High specific gravity, soluble with effervescence in dilute hydrochloric acid, colors the flame crimson. **Occurrence** Strontianite occurs in low-temperature hydrothermal veins, often in limestone, together with celestine, barite and calcite. It is a source of strontium and is used in fireworks and red flares. It is named after the locality of Strontian in Argyllshire (Highland), Scotland, where it was first found.

Witherite: pseudo-hexagonal twin

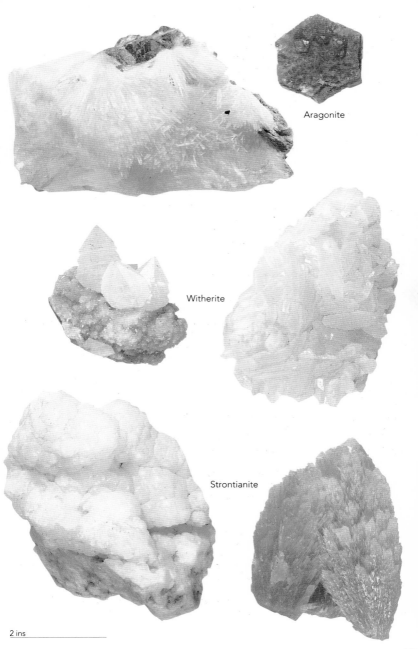

Aragonite

Witherite

Strontianite

2 ins

2 ins

Cerussite

Cerussite

Cerussite: star-shaped twin

Cerussite PbCO₃

Crystal system Orthorhombic. **Habit** Crystals often prismatic, or tabular parallel to side pinacoid; sometimes bipyramidal or pseudo-hexagonal star-like twins. Also acicular; or granular, massive, compact. **Twinning** Very common; either contact or penetration twins often of arrow-head shape. **SG** 6.4–6.6 **Hardness** 3–3½ **Cleavage** Prismatic in two directions, distinct. **Fracture** Conchoidal. **Color and transparency** Usually white or gray; sometimes darker colors: transparent to translucent. **Streak** White. **Luster** Adamantine. **Distinguishing features** High specific gravity, adamantine luster, dissolves with effervescence in warm dilute nitric acid, which distinguishes it from anglesite. **Occurrence** Cerussite is usually of secondary origin, occurring in the oxidized zone of lead veins. It is frequently found in association with anglesite, galena, smithsonite, pyromorphite and sphalerite. Cerussite is a lead ore, and its name comes from the Latin for "white lead".

Malachite Cu₂CO₃(OH)₂

Crystal system Monoclinic. **Habit** Crystals rare. Generally as botryoidal encrusting masses, in bands of varying color, and frequently of fibrous radiating habit. Also granular or earthy. **Twinning** Common. **SG** 3.9–4.0 (massive varieties as low as 3.5) **Hardness** 3½–4 **Cleavage** Pinacoidal, perfect. **Fracture** Subconchoidal, uneven. **Color and transparency** Bright green: translucent. **Streak** Pale green. **Luster** Fibrous varieties silky; rather dull when massive; crystals adamantine. **Distinguishing features** Color, botryoidal form, soluble with effervescence in dilute hydrochloric acid. **Occurrence** Malachite is a common secondary copper mineral, occurring typically in the oxidized zone of copper deposits. It is frequently associated with azurite, native copper and cuprite, which it sometimes replaces. Other accompanying minerals are calcite, chrysocolla and limonite.

Azurite (Chessylite) $Cu_3(CO_3)_2(OH)_2$
Crystal system Monoclinic. **Habit** Crystals often either tabular or short prismatic; as radiating aggregates; also massive or earthy. **SG** 3.8–3.9 **Hardness** $3\frac{1}{2}$–4 **Cleavage** Prismatic, perfect; pinacoidal, less so. **Fracture** Conchoidal. **Color and transparency** Various shades of deep azure-blue: transparent to translucent. **Streak** Light blue. **Luster** Vitreous. **Distinguishing features** Color, soluble with effervescence in nitric or hydrochloric acid. **Occurrence** Azurite, like malachite, is a secondary copper mineral and occurs with it in the oxidized zone of copper deposits. Azurite is not as widely distributed as malachite, though it is sometimes interbanded with it when massive. Pseudomorphs of malachite after azurite are common. Azurite often forms good, sharp crystals in contrast to malachite.

Azurite

2 ins

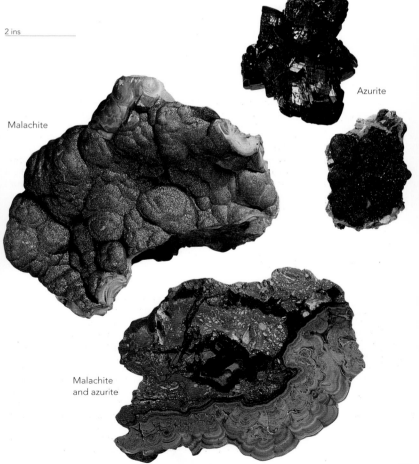

Azurite

Malachite

Malachite
and azurite

73

Nitratine (Chile saltpetre, Soda nitre) $NaNO_3$

Crystal system Trigonal. **Habit** Crystals rare, rhombohedral; usually massive. **Twinning** Common. **SG** 2.2–2.3 **Hardness** 1–2 **Cleavage** Rhombohedral, perfect. **Fracture** Conchoidal; rarely seen owing to perfect cleavage. **Color and transparency** Colorless or white; may be darker owing to impurities: transparent. **Streak** White. **Luster** Vitreous. **Distinguishing features** Low specific gravity and hardness, fuses easily, soluble in water, deliquescent. Nitratine resembles calcite but is lighter in weight and less hard. **Occurrence** Because of its ready solubility nitratine occurs in arid regions as surface deposits associated with gypsum, halite and other soluble nitrates and sulfates. Nitre, KNO_3, occurs together with nitratine under similar conditions but is less common. Nitratine is worked as a source of nitrate.

Borax

Borax $Na_2B_4O_5(OH)_4.8H_2O$

Crystal system Monoclinic. **Habit** Crystals prismatic; also massive. **SG** 1.7 **Hardness** $2-2\frac{1}{2}$ **Cleavage** Pinacoidal, perfect. **Fracture** Conchoidal. **Color and transparency** Colorless or white, sometimes grayish or tinged with blue: translucent. **Streak** White. **Luster** Vitreous to resinous, sometimes dull. **Distinguishing features** Crystal form, low specific gravity, soluble in water, fuses easily. **Occurrence** Borax is an evaporite mineral, precipitated by the evaporation of the water of saline lakes. It occurs in association with other evaporite minerals such as halite, sulfates, carbonates and other borates, in dried lakes in arid regions.

Colemanite $Ca_2B_6O_{11}.5H_2O$

Crystal system Monoclinic. **Habit** Crystals variable in habit, but usually short prismatic; also massive, compact, granular. **SG** 2.4 **Hardness** $4-4\frac{1}{2}$ **Cleavage** One perfect cleavage. **Fracture** Uneven. **Color and transparency** Colorless to white, also yellowish or gray: transparent to translucent. **Luster** Vitreous. **Distinguishing features** Crystal form, perfect cleavage, fuses easily, relatively hard for a borate. **Occurrence** Colemanite occurs in association with borax, but principally as a lining to cavities in sedimentary rocks where it was probably deposited from waters passing through primary borates. It is named after W. T. Coleman, a Californian industrialist.

Ulexite $NaCaB_5O_6(OH)_6.5H_2O$

Crystal system Triclinic. **Habit** Usually as rounded masses of fine fibrous crystals (cotton balls) and as parallel fibrous aggregates. **SG** 1.9–2.0 **Hardness** $2\frac{1}{2}$ (aggregates have an apparent hardness of 1) **Color and transparency** White: transparent. **Luster** Silky. **Distinguishing features** Soft "cotton ball" habit, low specific gravity, insoluble in cold water, slightly soluble in hot, fuses easily. **Occurrence** Ulexite is an evaporite mineral that sometimes accompanies colemanite in geodes in sedimentary rocks in areas of borax deposits. It occurs also with borax in the surface deposits of arid areas. It is named after G. L. Ulex, a 19th-century German chemist who discovered the mineral.

2 ins

Borax

Nitratine

Colemanite

Ulexite

2 ins

Barite

Barite: tabular habit

Barite: cockscomb mass

Barite (Barytes, Baryte) $BaSO_4$
Crystal system Orthorhombic. **Habit** Crystals commonly tabular, sometimes prismatic giving a diamond-shaped outline; also fibrous or lamellar, and in cockscomb masses; also granular and stalagmitic. **SG** 4.3–4.6 **Hardness** $2\frac{1}{2}$–$3\frac{1}{2}$ **Cleavage** Basal, perfect; prismatic, very good. **Fracture** Uneven. **Color and transparency** Colorless to white, often tinged with yellow, brown, blue, green or red: transparent to translucent. **Streak** White. **Luster** Vitreous. **Distinguishing features** High specific gravity, cleavage, crystal form, insoluble in acids, colors the flame green. **Occurrence** Barite is the most common mineral of barium. It occurs as a vein filling and as a gangue mineral accompanying ores of lead, copper, zinc, silver, iron, and nickel, together with calcite, quartz, fluorite, dolomite, and siderite. Barite also occurs as a replacement deposit of limestone, and as the cement in certain sandstones. Barite concretions in some sandstones have a characteristic rosette-like form and are called "desert roses". The name comes from a Greek word meaning "heavy".

Celestine (Celestite) SrSO$_4$

Crystal system Orthorhombic. **Habit** Crystals tabular or prismatic, resembling barite; also fibrous or granular. **SG** 3.9–4.0 **Hardness** 3–3$\frac{1}{2}$ **Cleavage** Basal, perfect; prismatic, good. **Fracture** Uneven. **Color and transparency** Colorless to faint bluish white; sometimes reddish: transparent to translucent. **Streak** White. **Luster** Vitreous. **Distinguishing features** High specific gravity, cleavage. Distinguished from barite, though often with difficulty, by lower specific gravity. As barium substitutes for strontium in the structure celestine grades into barite, but intermediate members are rare. Colors the flame crimson. **Occurrence** Celestine occurs in sedimentary rocks, particularly dolostone, as cavity linings associated with barite, gypsum, halite, anhydrite, calcite, dolomite, and fluorite. It occurs along with anhydrite in evaporite deposits. It is often associated with sulfur, both in the sedimentary environment and also in volcanic areas. It also occurs as a gangue mineral in hydrothermal veins with galena and sphalerite, and it forms concretionary masses in clay and marl. The name is taken from the Latin *caelestis*, meaning celestial, and alludes to the pale blue color of many crystals.

Barite: "desert rose"

Celestine: prismatic habit

Celestine

2 ins

Anglesite

Anglesite PbSO$_4$

Crystal system Orthorhombic. **Habit** Crystals sometimes tabular, often prismatic or pyramidal; also massive, compact, granular. **SG** 6.2–6.4 **Hardness** 2$\frac{1}{2}$–3 **Cleavage** Basal, good; prismatic, distinct. **Fracture** Conchoidal. **Color and transparency** Colorless to white, sometimes with a yellow, gray or bluish tinge: transparent to translucent. **Streak** White. **Luster** Adamantine. **Distinguishing features** High specific gravity (higher than barite), luster, association with galena. Distinguished from cerussite by lack of reaction with warm dilute nitric acid. **Occurrence** Anglesite is a secondary lead mineral, most commonly occurring in the oxidized zone of lead deposits; masses of anglesite often surround a core of galena.

Anhydrite CaSO$_4$

Crystal system Orthorhombic. **Habit** Crystals rare; usually massive, granular, fibrous. **SG** 2.9–3.0 **Hardness** 3–3$\frac{1}{2}$ **Cleavage** Three good cleavages, at right-angles. **Fracture** Uneven. **Color and transparency** Colorless to white, frequently with a bluish tinge, sometimes gray or reddish: transparent to translucent. **Streak** White. **Luster** Vitreous to pearly. **Distinguishing features** Three cleavages at right-angles, harder than gypsum, higher specific gravity than calcite. **Occurrence** Anhydrite is an evaporite mineral which occurs with gypsum and halite. It is deposited directly from sea water at temperatures in excess of 108°F, or it may form by the dehydration of gypsum. It occurs also in the "cap rock" above salt domes, and as a minor gangue mineral in hydrothermal metallic ore veins.

Anhydrite

2 ins

Gypsum

2 ins

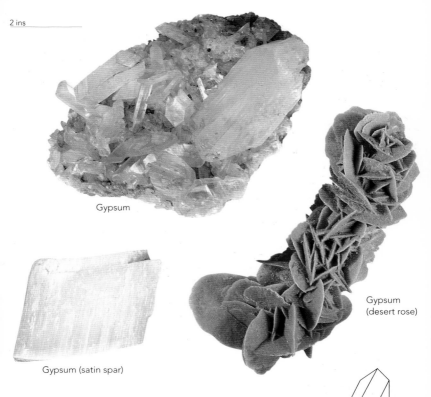

Gypsum

Gypsum
(desert rose)

Gypsum (satin spar)

Gypsum CaSO₄.2H₂O

Crystal system Monoclinic. **Habit** Crystals tabular, often with curved faces. The colorless, transparent variety is called selenite. Also fibrous (satin spar), massive, granular. The fine-grained granular variety is called alabaster. **Twinning** Very common, giving swallow-tail contact twins. **SG** 2.3 **Hardness** 2 **Cleavage** One perfect; two others, good. **Color and transparency** Colorless to white but sometimes in shades of yellow, gray, red and brown: transparent to translucent. **Streak** White. **Luster** Vitreous, pearly parallel to cleavage. **Distinguishing features** Low hardness (can be scratched with the finger nail), cleavage. **Occurrence** Gypsum is an evaporite mineral, and so occurs in bedded deposits together with halite and anhydrite. Having a low solubility, it is the first mineral to be precipitated from evaporating sea water, followed by anhydrite and then halite. It occurs in much smaller quantities in volcanic areas where sulfuric acid fumes have reacted with limestone, and in mineral veins where sulfuric acid produced by the oxidation of pyrite has reacted with calcareous wall rocks. Much gypsum is produced by the secondary hydration of anhydrite.

Gypsum

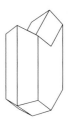

Gypsum: twinned crystal

79

2 ins

Chalcanthite

Epsomite

Chalcanthite

Chalcanthite

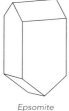

Epsomite

Chalcanthite $CuSO_4.5H_2O$
Crystal system Triclinic. **Habit** Crystals usually stout, prismatic; also massive, stalactitic, or fibrous. **SG** 2.1–2.3 **Hardness** $2\frac{1}{2}$ **Cleavage** Pinacoidal, imperfect. **Fracture** Conchoidal. **Color and transparency** Deep sky-blue: transparent to translucent. **Streak** White. **Luster** Vitreous. **Distinguishing features** Color, soluble in water. **Occurrence** Chalcanthite is rare, and is a secondary mineral of copper found usually in the oxidized zone of copper sulfide ore deposits. Because of its ready solubility, it is most commonly preserved in arid regions. It is also deposited from mine waters, having been recorded coating the walls of abandoned mines.

Epsomite (Epsom salt) $MgSO_4.7H_2O$
Crystal system Orthorhombic. **Habit** Natural crystals rare, crystals can be grown artificially; usually forms botryoidal encrusting masses with a fibrous structure. **SG** 1.7 **Hardness** $2–2\frac{1}{2}$ **Cleavage** One perfect cleavage. **Fracture** Conchoidal. **Color and transparency** Colorless to white: transparent to translucent. **Streak** White. **Luster** Vitreous; fibrous varieties silky to earthy. **Distinguishing features** Fibrous habit, readily soluble in water, bitter taste. **Occurrence** Epsomite usually occurs as encrusting masses on the walls of caves or mine workings where rocks rich in magnesium are exposed. It also occurs in the oxidized zone of pyrite deposits in arid regions.

Alunite

Alunite

Jarosite

2 ins

Alunite (Alumstone) $KAl_3(SO_4)_2(OH)_6$

Crystal system Trigonal. **Habit** Crystals rare; rhombohedral and pseudo-cubic; usually massive. **SG** 2.6–2.8 **Hardness** $3\frac{1}{2}$–4 **Cleavage** Basal, distinct. **Fracture** Uneven, conchoidal. **Color and transparency** White, sometimes gray or reddish: transparent to translucent. **Streak** White. **Luster** Vitreous; pearly parallel to cleavage. **Distinguishing features** Difficult to distinguish from massive dolomite, anhydrite or magnesite without chemical tests. Somewhat astringent taste. **Occurrence** Alunite is usually found as a secondary mineral in areas where volcanic rocks containing potassic feldspars have been altered by solutions containing sulfuric acid.

Jarosite $KFe_3(SO_4)_2(OH)_6$

Crystal system Trigonal. **Habit** Crystals minute pseudo-cubic rhombohedra; also fibrous, massive, granular, encrusting or nodular; often earthy. **SG** 3.2 **Hardness** $2\frac{1}{2}$–$3\frac{1}{2}$ **Cleavage** Basal, distinct. **Fracture** Uneven. **Color** Yellow ocher to dark brown. **Streak** Yellow. **Luster** Vitreous. **Distinguishing features** Color. **Occurrence** Jarosite forms under similar conditions to alunite, particularly where rocks contain ferric iron, and commonly in association with decomposing pyrite. It is most commonly found in volcanic areas around volcanic gas vents. It is named after Barranco Jaroso, a locality in Spain.

Thenardite

Glauberite: tabular habit

Crocoite

Crocoite

Thenardite Na₂SO₄

Crystal system Orthorhombic. **Habit** Crystals prismatic, tabular or pyramidal. **SG** 2.7 **Hardness** 2–3 **Cleavage** One perfect cleavage. **Color** White to brownish white. **Streak** White. **Occurrence** Thenardite is a rare evaporite mineral found with borates in evaporated salt lakes. It is named after L. J. Thénard (1777–1857), a French chemist.

Glauberite Na₂Ca(SO₄)₂

Crystal system Monoclinic. **Habit** Crystals prismatic or tabular. **SG** 2.7–2.8 **Hardness** 2½–3 **Cleavage** Basal, perfect. **Fracture** Conchoidal. **Color and transparency** Pale yellow to gray: transparent to translucent. **Streak** White. **Luster** Vitreous. **Distinguishing features** Thin tabular crystals, soluble in hydrochloric acid, partially soluble in water with loss of transparency. **Occurrence** Glauberite is an evaporite mineral that occurs in bedded salt deposits, in association with halite, thenardite, and polyhalite.

Polyhalite K₂Ca₂Mg(SO₄)₄.2H₂O

Crystal system Triclinic. **Habit** Crystals rare; usually as fibrous or lamellar masses. **Twinning** Common. **SG** 2.8 **Hardness** 2½–3 **Cleavage** Pinacoidal, distinct. **Color and transparency** Flesh-pink to brick-red: translucent. **Luster** Silky in fibrous masses, otherwise resinous. **Distinguishing features** Pink color, bitter taste. **Occurrence** Polyhalite occurs with glauberite in bedded evaporite deposits. It is one of the last minerals to be precipitated from saline waters owing to its high solubility.

Crocoite PbCrO₄

Crystal system Monoclinic. **Habit** Crystals usually prismatic or acicular, sometimes short prismatic; also massive, granular. **SG** 5.9–6.1 **Hardness** 2½–3 **Cleavage** Prismatic, distinct. **Fracture** Uneven. **Color and transparency** Orange-red to various shades of brown: translucent. **Streak** Orange-yellow. **Luster** Adamantine to vitreous. **Distinguishing features** Orange-red color, luster, high specific gravity, fuses fairly easily. **Occurrence** Crocoite is a rare secondary mineral that occurs in the oxidized zone of lead mineral veins together with other secondary lead minerals such as cerussite and pyromorphite. Chromium was first discovered in crocoite, and the name comes from the Greek word for "saffron".

Linarite PbCuSO₄(OH)₂

Crystal system Monoclinic. **Habit** Crystals prismatic. **SG** 5.3–5.4 **Hardness** 2½–3 **Cleavage** Pinacoidal, perfect; basal, distinct. **Color and transparency** Deep blue: translucent. **Luster** Vitreous. **Distinguishing features** Color, cleavage, association. Distinguished from azurite by lack of effervescence in dilute hydrochloric acid; instead, a white coating is developed. **Occurrence** Linarite is a rare but colorful secondary mineral that occurs in association with some lead-copper ores. The name comes from Linares, a locality in Spain.

2 ins

2 ins

Glauberite

Polyhalite

Crocoite

Linarite

Wolframite

Wulfenite: tabular crystal

Ferberite – Hübnerite series (Fe,Mn)WO$_4$

Crystal system Monoclinic. **Habit** Crystals tabular or prismatic. Often forms bladed, subparallel groups; also massive, granular. **Twinning** Contact twins occur. **SG** 7.0–7.5 **Hardness** 5–5½ **Cleavage** One perfect cleavage. **Fracture** Uneven. **Color and transparency** Gray-black to brownish black: opaque. **Streak** Brownish black. **Luster** Submetallic. **Distinguishing features** Color, one good cleavage, high specific gravity. There is virtually a complete series from ferberite (FeWO$_4$) to hübnerite (MnWO$_4$). Wolframite is an intermediate member of the series. The name was formerly used for the series as a whole. **Alteration** Sometimes alters to scheelite. **Occurrence** Wolframite occurs in quartz veins and pegmatites associated with granitic rocks, and is often accompanied by minerals such as cassiterite, arsenopyrite, tourmaline, scheelite, galena, sphalerite, and quartz. It is found also in high-temperature hydrothermal veins in association with the minerals listed above. Being heavy, it also occurs in some alluvial deposits.

Wolframite

2 ins

Scheelite

84

Scheelite CaWO$_4$
Crystal system Tetragonal. **Habit** Crystals usually bipyramidal; also massive, granular. **Twinning** Penetration twins are common. **SG** 5.9–6.1 **Hardness** 4½–5 **Cleavage** Pyramidal, distinct. **Color and transparency** White, sometimes in shades of yellow, green, brown, red: transparent to translucent. **Streak** White. **Luster** Vitreous. **Distinguishing features** Pyramidal habit, white color together with high specific gravity. Scheelite is commonly fluorescent. **Occurrence** Scheelite often accompanies wolframite in pegmatites and high-temperature hydrothermal veins. Associated minerals are cassiterite, molybdenite, fluorite, and topaz. It also occurs in contact metamorphic deposits together with vesuvianite, axinite, garnet, and wollastonite. Scheelite, with wolframite, is an ore of tungsten. It is named after K. W. Scheele, the 18th-century Swedish chemist who discovered tungsten.

Scheelite

Wulfenite PbMoO$_4$
Crystal system Tetragonal. **Habit** Crystals usually square plates or tablets; sometimes massive, granular. **SG** 6.5–7.0 **Hardness** 3 **Cleavage** Pyramidal, distinct. **Fracture** Subconchoidal. **Color and transparency** Orange-yellow, olive-green or brown, sometimes grayish: transparent to subtranslucent. **Streak** White. **Luster** Resinous to adamantine. **Distinguishing features** Orange-yellow color (commonly), luster, square tabular habit. **Occurrence** Wulfenite is a secondary mineral formed in the oxidized zone of ore deposits containing minerals of lead and molybdenum. It is commonly associated with anglesite, cerussite, vanadinite, and pyromorphite. It is named after F. X. Wülfen (1728–1805), an Austrian mineralogist.

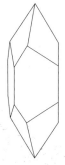

Wulfenite: bipyramidal habit

2 ins

Wulfenite

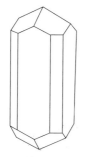

Xenotime: prismatic habit

Monazite

Xenotime YPO$_4$

Crystal system Tetragonal. **Habit** Crystals prismatic, resembling zircon with which it is sometimes associated in parallel growth. **SG** 4.4–5.1 **Hardness** 4–5 **Cleavage** Prismatic, perfect. **Fracture** Uneven. **Color and transparency** Yellowish brown, also grayish white, pale yellow: translucent to opaque. **Streak** Pale brown. **Luster** Resinous to vitreous. **Distinguishing features** Very similar to zircon but less hard and has a good prismatic cleavage. **Occurrence** Xenotime is an accessory mineral in granitic and alkaline igneous rocks. It also occurs in some pegmatites and gneisses.

Monazite (Ce,La,Nd,Th)PO$_4$

Crystal system Monoclinic. **Habit** Crystals small, short prismatic or tabular. Large crystals often have striated faces. **Twinning** Common. **SG** 4.9–5.4 **Hardness** 5–5$\frac{1}{2}$ **Cleavage** Pinacoidal, distinct. **Fracture** Uneven. **Color and transparency** Clove-brown to reddish brown, sometimes green: translucent. **Streak** Off-white. **Luster** Resinous to waxy. **Distinguishing features** Monazite-(Ce) is the species in which Ce predominates; monazite-(La) has La>Ce; and monazite-(Nd) has dominant Nd. Similar to zircon but softer. **Occurrence** Monazite is an accessory mineral of granitic rocks and associated pegmatites, and it also occurs in gneisses and carbonatites. It is concentrated in some detrital sands in sufficient quantity to merit commercial exploitation for cerium and thorium.

2 ins

Xenotime

Vivianite

Monazite

Amblygonite

2 ins

Vivianite

Vivianite $Fe^{2+}_3(PO_4)_2.8H_2O$
Crystal system Monoclinic. **Habit** Crystals prismatic; also as reniform or encrusting masses, often with a fibrous structure; sometimes blue, earthy. **SG** 2.6–2.7 **Hardness** $1\frac{1}{2}$–2 **Cleavage** One perfect cleavage. **Color** Colorless when fresh and unaltered, becoming blue or green on oxidation, the color deepening with exposure. **Streak** White, becoming dark blue or brown. **Luster** Vitreous, pearly parallel to cleavage. **Distinguishing features** Blue color. **Occurrence** Vivianite is a secondary phosphate which occurs in the oxidized zones of metallic ore deposits containing pyrrhotite and pyrite, in the weathered zone of certain phosphate-bearing pegmatites, and in sedimentary rocks, particularly those containing bone or other organic fragments.

Amblygonite - montebrasite $(Li,Na)AlPO_4(F,OH)$
Crystal system Triclinic. **Habit** Crystals usually rough and ill-formed, sometimes large; also massive, compact, or as cleavable masses. **Twinning** Lamellar twinning common. **SG** 3.0–3.1 **Hardness** $5\frac{1}{2}$–6 **Cleavage** Two good cleavages. **Fracture** Uneven. **Color and transparency** White to pale green or bluish white; sometimes pinkish or yellowish: subtransparent to translucent. **Streak** White. **Luster** Vitreous to greasy; pearly parallel to best cleavage. **Distinguishing features** Two cleavages, specific gravity. The name amblygonite is used for the fluorine-rich end member of the series, and montebrasite for the more common hydroxyl-rich member. **Occurrence** Amblygonite is a rare mineral which occurs in granite pegmatites together with other lithium minerals such as spodumene, tourmaline, and lepidolite, and with albite, for which it may be mistaken.

87

Apatite

Pyromorphite: mimetite, variety campylite, showing barrel-shaped form

Apatite Ca$_5$(PO$_4$)$_3$(F,Cl,OH)

Crystal system Hexagonal. **Habit** Crystals common, and are usually prismatic or tabular; also massive, granular. **SG** 3.1–3.3 **Hardness** 5 **Cleavage** Basal, imperfect. **Fracture** Conchoidal, uneven. **Color and transparency** Usually in shades of green to gray-green; also white, brown, yellow, bluish or reddish: transparent to translucent. **Streak** White. **Luster** Vitreous to subresinous. **Distinguishing features** Hexagonal crystal form, hardness. Distinguished from beryl, for which it may be mistaken, by inferior hardness; apatite can be scratched with a steel knife blade. Apatite is the group name: common, F-rich apatite is called fluorapatite. Pyromorphite, mimetite and vanadinite are also members of the group. **Occurrence** Apatite is a widely distributed phosphate mineral. It occurs as small crystals as an accessory mineral in a wide range of igneous rocks. Large crystals occur in pegmatites and in some high-temperature hydrothermal veins. It occurs also in both regional and contact metamorphic rocks, especially in metamorphosed limestones and skarns. In sedimentary rocks, apatite is a principal constituent of fossil bones and other organic matter. The name collophane is sometimes used for such phosphatic material. The name comes from a Greek word meaning "to deceive" because apatite, particularly the gem variety, is readily mistaken for other minerals.

Pyromorphite Pb$_5$(PO$_4$)$_3$Cl

Crystal system Hexagonal. **Habit** Crystals are usually of simple prismatic form, often barrel-shaped (campylite) or as hollow prismatic forms. Also fibrous, granular, globular. **SG** 6.5–7.1 **Hardness** $3\frac{1}{2}$–4 **Fracture** Subconchoidal. **Color and transparency** Shades of green, yellow and brown: subtransparent to translucent. **Streak** White. **Luster** Resinous. **Distinguishing features** Color, hexagonal form, high specific gravity, resinous luster. **Occurrence** Pyromorphite is a secondary lead phosphate that occurs, often with mimetite, in the oxidized zone of mineral veins containing lead minerals, such as galena and anglesite.

Mimetite Pb$_5$(AsO$_4$)$_3$Cl

Crystal system Hexagonal. **Habit** Crystals, like those of pyromorphite, are commonly simple hexagonal forms of prismatic habit; also as rounded, globular forms (campylite). **SG** 7.0–7.2 **Hardness** $3\frac{1}{2}$–4 **Fracture** Subconchoidal. **Color and transparency** Pale yellow to yellow-brown: subtransparent to translucent. **Streak** White. **Luster** Resinous. **Distinguishing features** Color, hexagonal form, high specific gravity, resinous luster. Mimetite and pyromorphite are often difficult to distinguish without chemical tests. **Occurrence** Mimetite is a rather rare secondary mineral that occurs in the oxidized parts of lead ores, especially those containing arsenic. Like pyromorphite, it occurs in association with galena, anglesite, and hemimorphite. The name comes from a Greek word meaning "imitator", because of the close resemblance between pyromorphite and mimetite.

2 ins

Pyromorphite

Apatite

Pyromorphite

Pyromorphite

Mimetite

Mimetite

Vanadinite

Vanadinite:
hollow prismatic crystals

Vanadinite $Pb_5(VO_4)_3Cl$

Crystal system Hexagonal. **Habit** Crystals frequently sharp and prismatic; sometimes as hollow prisms; also as rounded forms similar to pyromorphite. **SG** 6.7–7.1 **Hardness** 3 **Fracture** Subconchoidal. **Color and transparency** Orange-red, brownish-red to yellow: transparent to subtranslucent. **Streak** White to yellowish. **Luster** Resinous. **Distinguishing features** Like pyromorphite and mimetite it has hexagonal form, resinous luster and high specific gravity, but it is distinguished from them by its orange-red color. **Occurrence** Vanadinite is a rare mineral and, like pyromorphite, it occurs in the oxidized zone of sulfide ore deposits carrying galena and other lead minerals.

Erythrite (Cobalt bloom) $Co_3(AsO_4)_2.8H_2O$: Annabergite (Nickel bloom) $Ni_3(AsO_4)_2.8H_2O$

Crystal system Monoclinic. **Habit** Crystals usually prismatic, often acicular; also as radiating groups and reniform masses with a columnar structure; also as powdery coatings. **SG** 3.0–3.1 **Hardness** $1\frac{1}{2}$–$2\frac{1}{2}$ **Cleavage** One perfect cleavage. **Color and transparency** Erythrite, crimson-red to pink, becoming paler with increasing nickel content; annabergite, apple-green: transparent to subtranslucent. **Streak** Erythrite, red, but paler than color noted above. **Luster** Adamantine to vitreous; pearly parallel to cleavage. **Distinguishing features**: Pink color (erythrite); green color (annabergite), association with cobalt and nickel minerals. **Occurrence** Erythrite and annabergite are secondary minerals produced by the surface oxidation of primary cobalt and nickel minerals. Their occurrence as pink or green powdery coatings gives rise to the names "cobalt bloom" and "nickel bloom".

Vanadinite

2 ins

2 ins

Annabergite

Erythrite

Scorodite

Turquoise

Turquoise $CuAl_6(PO_4)_4(OH)_8.4H_2O$

Crystal system Triclinic. **Habit** Crystals rare and minute; usually massive, granular to cryptocrystalline as reniform or encrusting masses, or in veins. **SG** 2.6–2.8 **Hardness** 5–6 **Fracture** Conchoidal. **Color and transparency** Sky-blue, blue-green to greenish gray: nearly opaque. **Streak** White or greenish. **Luster** Waxy when massive; crystals vitreous. **Distinguishing features** Blue color, distinguished from chrysocolla by its greater hardness. **Occurrence** Turquoise is a secondary mineral occurring in veins in association with aluminous igneous or sedimentary rocks that have undergone considerable alteration, usually in arid regions. Prized as a semi-precious stone.

Scorodite $Fe^{3+}AsO_4.2H_2O$

Crystal system Orthorhombic. **Habit** Crystals pyramidal and pseudo-octahedral or prismatic; also nodular or earthy. **SG** 3.1–3.3 **Hardness** $3\frac{1}{2}$–4 **Cleavage** Prismatic, imperfect. **Fracture** Uneven. **Color and transparency** Pale green, blue-green to blue, brown: subtransparent to translucent. **Streak** White. **Luster** Vitreous to adamantine. **Distinguishing features** Crystal habit and association with arsenic minerals. **Occurrence** Scorodite is an alteration product of arsenic minerals, and of arsenopyrite in particular. It is also deposited from the waters of some hot springs. The name comes from the Greek word for "garlic" and refers to the odor it emits when heated.

Scorodite

2 ins

Torbernite

Autunite

Torbernite: scaly aggregate

Torbernite (and Metatorbernite) $Cu^{2+}(UO_2)_2(PO_4)_2.8-12H_2O$
Crystal system Tetragonal. **Habit** Crystals tabular, often with square outline; also as foliated or scaly aggregates. **SG** 3.2 (increasing to 3.7 with alteration to metatorbernite) **Hardness** $2-2\frac{1}{2}$ **Cleavage** Basal, perfect. **Color and transparency** Bright emerald-green, sometimes dark green: transparent to translucent. **Streak** Pale green. **Luster** Vitreous; pearly parallel to cleavage. **Distinguishing features** Color, cleavage. **Occurrence** Torbernite and autunite occur together as secondary minerals in the oxidized parts of veins containing uraninite and copper minerals. At atmospheric temperatures, torbernite loses some of its water and forms metatorbernite $Cu(UO_2)_2(PO_4)_2.8H_2O$.

Autunite (and Meta-autunite) $Ca(UO_2)_2(PO_4)_2.10-12H_2O$
Crystal system Tetragonal. **Habit** Tabular crystals with square outline, very similar in shape to torbernite; also as foliated and scaly masses. **SG** 3.1–3.2 **Hardness** $2-2\frac{1}{2}$ **Cleavage** Basal, perfect. **Color** Lemon-yellow to greenish yellow. **Streak** Yellow. **Luster** Vitreous; pearly parallel to cleavage. **Distinguishing features** Yellow-green color, cleavage, crystal form. Easily distinguished from torbernite by color, but autunite can be distinguished from other secondary uranium minerals only by chemical or X-ray methods. Fluorescent. **Occurrence** Autunite, like torbernite, is a secondary mineral that occurs in the oxidized parts of veins and pegmatites carrying uranium minerals. Autunite may lose some of its water forming meta-autunite $Ca(UO_2)_2(PO_4)_2.2-6H_2O$.

Tyuyamunite

Carnotite

2 ins

Carnotite $K_2(UO_2)_2(VO_4)_2.3H_2O$

Crystal system Monoclinic. **Habit** Usually powdery; rarely as minute, thin tabular crystals. **SG** 4–5 **Hardness** About 2 **Cleavage** Basal, perfect. **Color** Bright yellow to greenish yellow. **Luster** Dull, earthy. **Distinguishing features** Yellow color, though it is difficult to distinguish from tyuyamunite other than by chemical or X-ray means. Its powdery habit and lack of fluorescence distinguish it from autunite. **Occurrence** Carnotite and tyuyamunite are secondary minerals frequently found together in sedimentary rocks, having been deposited from waters which have been in contact with primary uranium and vanadium minerals.

Tyuyamunite (and Metatyuyamunite)
$Ca(UO_2)_2(VO_4)_2.5–8H_2O$

Crystal system Orthorhombic. **Habit** Scales, laths or radial aggregates; also massive or powdery. **SG** 3.6 (increasing to 4.4 with alteration to metatyuyamunite) **Hardness** $2–2\frac{1}{2}$ **Cleavage** Basal, perfect. **Color** Greenish yellow. **Luster** Earthy; massive material waxy. **Distinguishing features** Closely resembles carnotite, but has more greenish color and is fluorescent. **Occurrence** Tyuyamunite, like carnotite, with which it is commonly associated, is a secondary mineral that occurs in sedimentary rocks, notably certain sandstones. Together with carnotite, it is an ore of uranium. Metatyuyamunite has 3–5 molecules of water per formula unit. The strange-sounding name is taken from Tyuya-Muyun, a locality in Uzbekistan.

Descloizite Pb(Zn,Cu)VO₄OH

Crystal system Orthorhombic. **Habit** Crystals platy, prismatic or wedge-shaped; also mammillary with fibrous radiating structure. **SG** 5.9–6.2 **Hardness** 3½ **Color and transparency** Usually clove-brown but varies from cherry-red to black: translucent. **Streak** Orange to brownish red. **Distinguishing features** Color, orange streak, crystal form. **Occurrence** Descloizite is a secondary mineral found occasionally in lead-zinc deposits.

Olivenite Cu₂AsO₄OH

Crystal system Monoclinic. **Habit** Crystals prismatic or acicular; also reniform, fibrous, radiating or granular. **SG** 4.1–4.4 **Hardness** 3 **Cleavage** Poor. **Fracture** Conchoidal to uneven. **Color and transparency** Olive-green in various shades (hence the name), but can range from white to nearly black: subtransparent to opaque. **Streak** Olive-green to brown. **Luster** Vitreous; some fibrous varieties, pearly. **Distinguishing features** Olive-green color. **Occurrence** Olivenite is a rare secondary mineral which occurs in the oxidized parts of copper sulfide deposits, sometimes in association with adamite.

Descloizite

Olivenite

2 ins

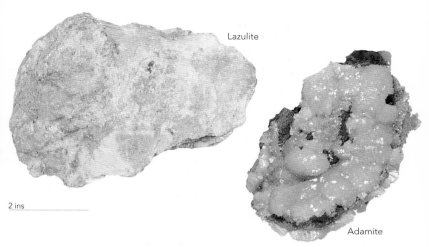

Lazulite

2 ins

Adamite

Adamite Zn₂AsO₄OH

Crystal system Orthorhombic. **Habit** Crystals usually small; more often as radiating and encrusting aggregates. **SG** 4.3–4.4 **Hardness** $3\frac{1}{2}$ **Color and transparency** Yellowish green to green, sometimes reddish brown: translucent. **Distinguishing features** Yellowish green color. **Occurrence** Adamite is a rare secondary zinc mineral found as a weathering product in the oxidized zone of zinc deposits.

Lazulite (Mg,Fe)Al₂(PO₄)₂(OH)₂

Crystal system Monoclinic. **Habit** Crystals sharp bipyramids; also massive, granular to compact. **SG** 3.0–3.1 **Hardness** 5–6 **Cleavage** Prismatic, indistinct. **Fracture** Uneven. **Color and transparency** Deep azure-blue: translucent. **Streak** White. **Luster** Vitreous. **Distinguishing features** Color, bipyramidal crystal form. When massive, lazulite is difficult to distinguish from other deep blue minerals. **Occurrence** Lazulite is a rare mineral which occurs in pegmatites and quartz veins and in quartzites. Associated minerals are kyanite, corundum, rutile, and sillimanite. Lazulite is used as a semi-precious stone, and the name comes from an Arabic word meaning "heaven" in allusion to the blue color.

Lazulite

Wavellite Al₃(PO₄)₂(OH,F)₃.5H₂O

Crystal system Orthorhombic. **Habit** Crystals rare; characteristically as hemispherical or globular aggregates with a fibrous, radiating structure. **SG** 2.3–2.4 **Hardness** $3\frac{1}{2}$–4 **Cleavage** Prismatic, good. **Fracture** Uneven. **Color and transparency** White; often greenish, yellow, gray, brown: translucent. **Streak** White. **Luster** Vitreous. **Distinguishing features** The radiating habit is the characteristic feature. **Occurrence** Wavellite is a secondary mineral found on joint surfaces and in cavities in rocks, particularly slates. It occurs in limonitic ore bodies and in association with phosphorite deposits. Wavellite is named after W. Wavell, the discoverer of the mineral.

Wavellite

95

Olivine group (Mg,Fe)$_2$SiO$_4$

Crystal system Orthorhombic. **Habit** Good crystals rare; occurs usually as isolated grains in igneous rocks, or as granular aggregates. **SG** 3.2–4.4 (increasing with iron content), common olivine about 3.3–3.4 **Hardness** 6$\frac{1}{2}$–7 **Cleavage** Pinacoidal, indistinct **Fracture** Conchoidal. **Color and transparency** Clear olivine-green (hence the name); sometimes yellowish or brownish to black; reddish when oxidized: transparent to translucent. **Luster** Vitreous. **Distinguishing features** Color, conchoidal fracture, association. The olivines range continuously in composition from forsterite (Mg$_2$SiO$_4$) to fayalite (Fe$_2$SiO$_4$), and some of the physical and optical properties vary with increasing content of iron. **Alteration** Olivine alters readily as a result of weathering or hydrothermal action. The usual alteration products are serpentine, iddingsite or saponite (bowlingite); iddingsite is a mixture of more than one mineral. **Occurrence** Olivine is a rock-forming mineral which occurs in silica-poor igneous rocks such as basalt, gabbro, troctolite, and peridotite. Dunite is a rock composed exclusively of olivine, and olivine nodules, composed mainly of olivine with some pyroxene, occur in some basalts. Olivine is produced also as a result of the metamorphism of magnesian sediments, particularly of siliceous dolostones, and in these rocks it is usually close to forsterite in composition. Fayalite occurs in certain rapidly cooled siliceous igneous rocks, such as pitchstone. In addition, olivine is a component of certain stony-iron meteorites, and it is abundant in lunar basalts.

Olivine

2 ins

Olivine

2 ins

Willemite

Monticellite

Willemite

Willemite Zn_2SiO_4

Crystal system Trigonal. **Habit** Crystals prismatic; usually massive, granular. **SG** 3.9–4.2 **Hardness** $5\frac{1}{2}$ **Cleavage** Basal, good. **Fracture** Uneven. **Color and transparency** Greenish yellow is typical; but varies from near white to brown: transparent to nearly opaque. **Luster** Vitreous to resinous. **Distinguishing features** Greenish color, association. Willemite is usually strongly fluorescent. **Occurrence** Willemite occurs in the oxidized zone of zinc ore deposits, but never in large amounts. It is named in honor of King William I (Willem Frederik) of the Netherlands.

Willemite

Monticellite $CaMgSiO_4$

Crystal system Orthorhombic. **Habit** Crystals small and prismatic; also as grains. **SG** 3.1–3.3 **Hardness** $5\frac{1}{2}$ **Color** Colorless to gray. **Occurrence** Monticellite occurs in metamorphosed calcareous rocks, usually impure dolostones. It is named after T. Monticelli (1759–1846), an Italian mineralogist.

*Phenakite:
rhombohedral habit*

Phenakite (Phenacite) Be_2SiO_4
Crystal system Trigonal. **Habit** Crystals often rhombohedral, sometimes prismatic. **SG** 3.0 **Hardness** $7\frac{1}{2}-8$ **Cleavage** Prismatic, poor. **Fracture** Conchoidal. **Color and transparency** Colorless, white, yellow, pinkish, brown: transparent to translucent. **Luster** Vitreous. **Distinguishing features** Crystal form, hardness. **Occurrence** Phenakite is a rare beryllium mineral that occurs in cavities in granites and in granite pegmatites in association with beryl, topaz, and apatite. It also occurs in metamorphic rocks carrying beryl and in hydrothermal veins. Phenakite is sometimes used as a gemstone, and the name comes from the Greek word meaning "deceiver" because of its close resemblance to quartz.

Phenakite

2 ins

Dioptase

Dioptase $CuSiO_2(OH)_2$

Crystal system Trigonal. **Habit** Crystals usually short prismatic, often terminated by rhombohedra; also massive. **SG** 3.3 **Hardness** 5 **Cleavage** Rhombohedral, perfect. **Fracture** Conchoidal to uneven. **Color and transparency** Emerald-green: transparent to translucent. **Luster** Vitreous. **Distinguishing features** Color, crystal form, association with copper minerals. **Occurrence** Dioptase is not a common mineral but is found in the oxidized parts of copper sulfide deposits.

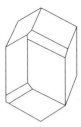

Dioptase

Humite group $Mg(OH,F)_2 \cdot 1{-}4(Mg,Fe^{2+})_2SiO_4$

Crystal system Orthorhombic and monoclinic. **Habit** Crystals usually of stubby but varied habit; also massive. **Twinning** Common. **SG** 3.1–3.3 **Hardness** $6{-}6\frac{1}{2}$ **Cleavage** One poor cleavage. **Fracture** Uneven. **Color and transparency** White, pale yellow, brown: translucent. **Luster** Vitreous to resinous. **Distinguishing features** Light yellow or brownish color, association with metamorphosed limestone. Colorless varieties difficult to distinguish from olivine. The humite group comprises four minerals, norbergite, chondrodite, humite, and clinohumite, which differ only in the amount of magnesia, iron, and silica they contain. Humite and norbergite are orthorhombic, and chondrodite and clinohumite monoclinic, but the monoclinic minerals depart only slightly from orthorhombic symmetry. Individual members of the series are difficult to distinguish in hand specimen. **Occurrence** Members of the humite group occur typically in metamorphosed dolomitic limestones and spinel, phlogopite, garnet, vesuvianite, diopside, graphite, and calcite are associated minerals. The name chondrodite is taken from a Greek word meaning "a grain". Humite is named after Sir Abraham Hume (1748–1838); and norbergite is named after Norberg, a locality in Sweden.

2 ins

Humite

Chondrodite

Zircon

Zircon: knee-shaped twin

Zircon ZrSiO$_4$

Crystal system Tetragonal. **Habit** Crystals usually prismatic, with bipyramidal terminations. **Twinning** Common, giving knee-shaped twins. **SG** 4.6–4.7 **Hardness** 7½ **Cleavage** Prismatic, indistinct. **Fracture** Conchoidal, very brittle. **Color and transparency** Variable, though most commonly light brown to reddish brown; also colorless, gray, yellow, green: transparent to translucent, occasionally nearly opaque. **Luster** Vitreous to adamantine. **Distinguishing features** Square prismatic habit, brownish color, hardness, high specific gravity. **Occurrence** Zircon is one of the most widely distributed accessory minerals in igneous rocks such as granite, syenite, and nepheline syenite. In pegmatites, the crystals sometimes reach a considerable size. It occurs also in metamorphic rocks such as schists and gneisses and, owing to its specific gravity and durability, it becomes concentrated as a detrital mineral in beach and river sands. Transparent zircon is used as a gemstone, and brownish varieties are called hyacinth or jacinth. It is a source of the metal zirconium, which took its name from the mineral. The name zircon is very old and may come from Persian words meaning "golden color".

2 ins

Zircon

Titanite

Dumortierite

2 ins

Eudialyte

Dumortierite

Titanite (Sphene) $CaTiSiO_5$
Crystal system Monoclinic. **Habit** Crystals commonly flattened and wedge-shaped; sometimes massive. **Twinning** Common. **SG** 3.4–3.6 **Hardness** 5–5½ **Cleavage** Prismatic distinct. **Fracture** Conchoidal. **Color and transparency** Brown and greenish yellow are the commonest colors, sometimes gray or nearly black: transparent to translucent, occasionally nearly opaque. **Luster** Resinous to adamantine. **Distinguishing features** Sharp, wedge-shaped habit, adamantine luster, greenish yellow color. **Occurrence** Titanite is widely distributed as an accessory mineral, particularly in coarse-grained igneous rocks such as syenite, nepheline syenite, diorite, and granodiorite. It occurs similarly in schists and gneisses and in some metamorphosed limestones. The name alludes to its composition; sphene comes from a Greek word meaning "wedge" and refers to the crystal habit.

Titanite

Dumortierite $Al_7BO_3(SiO_4)_3O_3$
Crystal system Orthorhombic. **Habit** Crystals rare; usually in fibrous radiating aggregates. **SG** 3.3–3.4 **Hardness** 7 **Cleavage** One poor cleavage. **Color and transparency** Bright greenish blue, violet, pink: transparent to translucent. **Luster** Vitreous. **Distinguishing features** Color, fibrous habit. **Occurrence** Dumortierite is a rare mineral that occurs in some schists, gneisses, and pegmatites. It is named after E. Dumortier, a French paleontologist.

Eudialyte $Na_{16}Ca_6(Fe^{2+},Mn^{2+},Y)_3Zr_3(Si_3O_9)_2(Si_9O_{27})_2(OH,Cl)_4$
Crystal system Trigonal. **Habit** Crystals rhombohedral or tabular; also massive, granular. **SG** 2.8–3.0 **Hardness** 5–5½ **Cleavage** Basal, indistinct. **Color and transparency** Red to brown: transparent to translucent. **Distinguishing features** Color, association with nepheline syenite. **Occurrence** Eudialyte occurs typically in nepheline syenites and nepheline syenite pegmatites.

2 ins

Almandine

Spessartine

Uvarovite

Almandine

Garnet group

General formula $X_3Y_2Si_3O_{12}$, where X is commonly Ca, Mn, Mg or Fe^{2+}; and Y is Al, Cr^{3+} or Fe^{3+}. Specific names are given to garnets of simple composition, though natural garnets rarely conform to such simple end members owing to substitution of one atom for another. The following names are in common use:

Garnet:
rhombic dodecahedron

Pyrope	$Mg_3Al_2Si_3O_{12}$
Almandine	$Fe_3^{2+}Al_2Si_3O_{12}$
Spessartine	$Mn_3Al_2Si_3O_{12}$
Grossular	$Ca_3Al_2Si_3O_{12}$
Uvarovite	$Ca_3Cr_2Si_3O_{12}$
Andradite	$Ca_3Fe_2^{3+}Si_3O_{12}$

There are, in effect, two main groups of garnets; the pyrope-almandine-spessartine group, and the grossular-uvarovite-andradite group. Continuous atomic substitutions take place within these groups, but there is no continuous substitution between them. **Crystal system** Cubic. **Habit** Crystals common; usually rhombdodecahedra or trapezohedra, or combinations of the two. Other forms occur but more rarely. Sometimes massive, granular. **SG** 3.6–4.3 (varying with composition) **Hardness** $6-7\frac{1}{2}$ **Cleavage** None. **Fracture** Subconchoidal. **Color and transparency** Color varies with composition: transparent to translucent. Pyrope, almandine, and spessartine are usually shades of deep red and

Garnet:
trapezohedron

2 ins

Grossular

Hessonite

Melanite

brown to nearly black; uvarovite is a clear green; grossular is brown, pale green or white; and andradite is yellow, brown or black. **Luster** Vitreous to resinous. **Distinguishing features** Hardness, rhombdodecahedral or trapezohedral crystal form. Individual members may be distinguished by color and specific gravity. However, chemical analysis is needed for precise determination. **Occurrence** Garnets are widely distributed in metamorphic and some igneous rocks. There is a link between composition and occurrence of garnets. Pyrope occurs in igneous rocks such as peridotite and in associated serpentinites, and also in kimberlite. Almandine is the common garnet of schists and gneisses; spessartine occurs in low grade metamorphic rocks, particularly if they contain manganese, and in some granites and pegmatites. Uvarovite is the rarest of the six garnet varieties listed here, occurring mainly in chromium-bearing serpentinites; grossular is characteristic of metamorphosed impure limestones; andradite occurs in metamorphosed limestones and in metasomatic calcareous rocks; and the black variety, called melanite, occurs in some feldspathoidal igneous rocks, such as phonolite and leucitophyre. Garnet is often a constituent of beach and river sands. Some varieties of garnet are used as gemstones. Hessonite (cinnamon stone) is yellow to brownish red and is a variety of grossular; demantoid is green andradite and is the best of the gem garnets; and rhodolite is a rose-colored or purplish garnet of the pyrope-almandine series.

Pyrope

Garnet: combination of rhombic dodecahedron and trapezohedron

103

Sillimanite

2 ins

Andalusite
(chiastolite)

Andalusite

Andalusite

*Andalusite
(chiastolite)*

Andalusite Al_2SiO_5

Crystal system Orthorhombic. **Habit** Crystals prismatic and pseudo-tetragonal with a square cross-section; also massive. Some crystals have carbonaceous inclusions arranged so that in cross-section they form a dark cross. This variety is called chiastolite. **SG** 3.1–3.2 **Hardness** $6\frac{1}{2}$–$7\frac{1}{2}$ **Cleavage** Prismatic, distinct. **Fracture** Uneven. **Color and transparency** Commonly pink or red; also gray, brown and green: transparent to nearly opaque. **Luster** Vitreous. **Distinguishing features** Square prismatic form, hardness, occurrence in metamorphic rocks. **Alteration** Andalusite alters to an aggregate of white mica flakes which often coat the crystals. **Occurrence** Andalusite occurs typically in thermally metamorphosed pelitic rocks, and in pelites that have been regionally metamorphosed under low-pressure conditions. It occurs also in some pegmatites, together with corundum, tourmaline, topaz, and other minerals. Transparent green andalusite is used as a gemstone. The name is derived from the Spanish province of Andalusia.

Sillimanite (Fibrolite) Al_2SiO_5

Crystal system Orthorhombic. **Habit** Commonly as elongated prismatic crystals, often fibrous and as felted (interwoven) masses. **SG** 3.2–3.3 **Hardness** $6\frac{1}{2}$–$7\frac{1}{2}$ **Cleavage** Pinacoidal, good. **Fracture** Uneven. **Color and transparency** Colorless, white, yellowish or brownish: transparent to translucent. **Luster** Vitreous. **Distinguishing features** Fibrous habit, though in this it resembles other fibrous silicates, from which it may be distinguished with the microscope, or by its association. **Occurrence** Sillimanite occurs typically in schists and gneisses produced by high-grade regional metamorphism. It is named after B. Silliman, an American chemist.

Kyanite (Disthene) Al₂SiO₅

Kyanite (Disthene) Al_2SiO_5

Crystal system Triclinic. **Habit** Crystals usually of flat, bladed habit, also as radiating bladed aggregates. **SG** 3.5–3.7 **Hardness** $5\frac{1}{2}$–7 (hardness is variable, being $5\frac{1}{2}$ along the length of the crystals and 6–7 across them) **Cleavage** Two good cleavages. **Color and transparency** Blue to white, but may be gray or green. Crystals are often unevenly colored, the darkest tints being at the centers of crystals, or as streaks and patches: transparent to translucent. **Luster** Vitreous, sometimes pearly on cleavage surfaces. **Distinguishing features** Blue color, bladed habit, good cleavage, variable hardness. **Occurrence** Kyanite occurs typically in regionally metamorphosed schists and gneisses, together with garnet, staurolite, mica, and quartz. It occurs also in pegmatites and quartz veins associated with schists and gneisses. The name kyanite comes from a Greek word meaning "blue": disthene alludes to the hardness that varies with direction in the mineral.

Kyanite: bladed crystal

Andalusite, sillimanite, and kyanite provide an example of polymorphism. There is a relationship between their structure (and hence specific gravity) and their mode of formation, the least dense andalusite forming under low-pressure metamorphism and kyanite, the densest, with a closely packed structure, forming under high-pressure conditions.

Kyanite

2 ins

Staurolite (Fe^{2+},Mg,Zn)$_2$Al$_9$(Si,Al)$_4$O$_{22}$(OH)$_2$

Crystal system Monoclinic, pseudo-orthorhombic. **Habit** Crystals usually prismatic; rarely massive. **Twinning** Common, giving rise to cruciform twins with the two individuals crossing either at right-angles or obliquely. **SG** 3.7–3.8 **Hardness** 7–7½ **Cleavage** One, distinct cleavage. **Fracture** Subconchoidal. **Color and transparency** Reddish brown to brown-black: translucent to nearly opaque. **Luster** Vitreous to resinous. **Distinguishing features** Brown color, crystal form (particularly if twinned). **Occurrence** Staurolite occurs typically as porphyroblasts in medium-grade schists and gneisses often in association with garnet, kyanite, and mica. The name comes from a Greek word meaning "cross" in allusion to the form of the twins.

Staurolite: cruciform twin

Staurolite: cruciform twin

2 ins

Staurolite

2 ins

Topaz

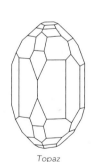

Euclase

Topaz $Al_2SiO_4(OH,F)_2$
Crystal system Orthorhombic. **Habit** Crystals usually prismatic, often with two or more vertical prism forms, or with striated prism faces. Also massive, granular. **SG** 3.5–3.6 **Hardness** 8 **Cleavage** Basal, perfect. **Fracture** Subconchoidal to uneven. **Color and transparency** Colorless; also pale yellow, pale blue, greenish and, rarely, pink: transparent to translucent. **Luster** Vitreous. **Distinguishing features** Crystal form, hardness, perfect basal cleavage, high specific gravity. **Occurrence** Topaz occurs typically in granite pegmatites, rhyolites and quartz veins. It also occurs as grains in granites which have been subjected to alteration by fluorine-bearing solutions, and is accompanied by fluorite, tourmaline, apatite, beryl, and cassiterite. Topaz also occurs as worn grains and pebbles in alluvial deposits. It is used as a gemstone.

Topaz

Euclase $BeAlSiO_4OH$
Crystal system Monoclinic. **Habit** Crystals prismatic. **SG** 3.0–3.1 **Hardness** $7\frac{1}{2}$ **Cleavage** One perfect cleavage, hence the name. **Color and transparency** Colorless to pale blue-green: transparent to translucent. **Luster** Vitreous. **Occurrence** Euclase is a rare mineral found in pegmatites in association with other beryllium minerals, notably beryl. It is sometimes used as a gemstone.

Epidote

Epidote group

Epidotes have the general formula $X_2Y_3Si_3O_{12}OH$, in which X is commonly Ca, and Y is usually Al and Fe^{3+}, partly replaced by Mg and Fe^{2+} in some species.

Zoisite $Ca_2Al_3Si_3O_{12}OH$

Crystal system Orthorhombic. **Habit** Crystals prismatic; also massive. **SG** 3.2–3.4 (increasing with iron content) **Hardness** 6 **Cleavage** One perfect cleavage. **Fracture** Uneven. **Color and transparency** Gray; sometimes pale green or brown. A pink, manganese-bearing variety is called thulite: transparent to sub-translucent. **Luster** Vitreous, pearly on cleavage surfaces. **Distinguishing features** Color, single perfect cleavage. **Occurrence** Zoisite occurs in schists and gneisses and in metasomatic rocks, together with garnet, vesuvianite, and actinolite. It occurs occasionally in hydrothermal veins. The blue variety called tanzanite is a valuable gemstone. Zoisite is named after Baron von Zois, an Austrian.

Clinozoisite $Ca_2Al_3Si_3O_{12}OH$ / **Epidote** (Pistacite) $Ca_2(Al,Fe^{3+})_3Si_3O_{12}OH$

Crystal system Monoclinic. **Habit** Crystals prismatic and often striated parallel to their length; also massive, fibrous or granular. **Twinning** Uncommon. **SG** 3.2–3.5 (increasing with iron content) **Hardness** 6–7 **Cleavage** One perfect cleavage, parallel to the length of the crystals. **Fracture** Uneven. **Color and transparency** Clinozoisite is usually greenish gray; epidote is yellowish green to black: transparent to nearly opaque. **Luster** Vitreous. **Distinguishing features** Distinctive yellow-green color, prismatic habit. Epidote can be mistaken for tourmaline, but the latter lacks cleavage and has a hexagonal or triangular cross-section. **Occurrence** Clinozoisite and epidote are widespread in medium- to low-grade metamorphic rocks, especially those derived from igneous rocks such as basalt and diabase, or from calcareous sediments. They also occur in contact metamorphosed limestones and in veins in igneous rocks.

2 ins

Epidote

Zoisite

Piemontite

Allanite

Zoisite (thulite)

Allanite (Orthite) $(Ca,Ce,Y,La,)_2(Al,Fe^{3+})_3Si_3O_{12}OH$
Crystal system Monoclinic. **Habit** Crystals prismatic, sometimes tabular; also massive. **SG** 3.4–4.2 **Hardness** $5–6\frac{1}{2}$ **Cleavage** Two poor cleavages. **Fracture** Conchoidal to uneven. **Color and transparency** Light brown to black: subtranslucent to opaque. **Luster** Vitreous or pitchy to submetallic. **Distinguishing features** Dark color, pitchy luster, weak radioactivity. Allanite-(Ce), allanite-(La) and allanite-(Y) are named according to the dominant rare-earth element present. **Occurrence** Allanite occurs as an accessory mineral in many granites, syenites, pegmatites, gneisses, and skarns. It is named after T. Allan (1777–1833), a British mineralogist.

2 ins

Piemontite (Piedmontite) $Ca_2(Al,Fe^{3+},Mn^{3+})_3Si_3O_{12}OH$
Crystal system Monoclinic. **Habit** As for epidote. **SG** 3.4–4.5 **Hardness** 6 **Cleavage** One perfect cleavage. **Fracture** Uneven. **Colour and transparency** Reddish or purplish brown to black. **Luster** Vitreous. **Distinguishing features** Reddish brown color. **Occurrence** Piemontite is a rare mineral which occurs in some low-grade schists and manganese ore deposits.

Axinite (Ca,Mn,Mg,Fe)$_3$Al$_2$BSiO$_4$O$_{15}$OH

Crystal system Triclinic. **Habit** Crystals commonly broad and with sharp edges; also massive, lamellar or granular. **SG** 3.3–3.4 **Hardness** $6\frac{1}{2}$–7 **Cleavage** One good cleavage. **Fracture** Conchoidal. **Color and transparency** Most crystals are a distinctive clove-brown color; also yellowish or gray: transparent to translucent. **Luster** Vitreous. **Distinguishing features** Color, sharp-edged crystal form. Axinite is a group name and derives from the Greek word for "axe", which is descriptive of the crystal form. **Occurrence** Axinite occurs in calcareots rocks that have undergone contact metamorphism and metasomatism, and in cavities in granites, especially close to their contacts.

Axinite

2 ins

Beryl (emerald)

Axinite

Beryl Be$_3$Al$_2$Si$_6$O$_{18}$

Crystal system Hexagonal. **Habit** Crystals usually prismatic often with striations parallel to their length; also massive. **SG** 2.6–2.8 **Hardness** $7\frac{1}{2}$–8 **Cleavage** Basal, poor. **Fracture** Conchoidal to uneven. **Color and transparency** Green, blue, yellow, pink but rather variable: transparent to translucent. Transparent, gem-quality beryl may be dark or light green (emerald), bluish green (aquamarine), yellow (heliodor) or pink (morganite). **Luster** Vitreous. **Distinguishing features** Hexagonal crystal form, green color (usually). It resembles apatite, but the greater hardness of beryl is distinctive. Massive beryl can be mistaken for quartz. **Occurrence** Beryl most commonly occurs as an accessory mineral in granites, and is usually found in cavities and in granite pegmatites. Beryl crystals in some pegmatites grow to very large sizes. It occurs also in mica schists and gneisses in association with phenakite, rutile, and chrysoberyl. Some varieties are prized as gemstones and beryl is a source of beryllium.

Beryl

Beryl (heliodor)

Beryl (aquamarine)

Beryl (emerald)

Beryl (aquamarine)

Beryl

Cordierite

Cordierite $(Mg,Fe)_2Al_4Si_5O_{18}$
Crystal system Orthorhombic. **Habit** Crystals prismatic and
pseudo-hexagonal but rather rare; usually as grains, or massive.
Twinning Common, giving pseudo-hexagonal forms. **SG** 2.5–2.8
(increasing with iron content) **Hardness** 7 **Cleavage** One poor
cleavage, basal parting. **Fracture** Subconchoidal to uneven.
Color and transparency Dark blue, grayish blue: transparent
to translucent. **Luster** Vitreous. **Distinguishing features** Dark blue
color but granular cordierite closely resembles quartz. The gem
variety, called iolite or dichroite, is deep blue or pale yellow
depending on the direction in which it is viewed. **Alteration** To an
aggregate of chlorite and muscovite called pinite. **Occurrence**
Cordierite occurs in aluminous rocks that have undergone
medium- to high-grade contact or regional metamorphism. It is
found in hornfelses, schists, and gneisses in association with
andalusite, spinel, quartz, and biotite. It also occurs in igneous
rocks which have assimilated aluminous sediments. Cordierite is
named after P. L. A. Cordier, a French geologist.

2 ins

111

Tourmaline

Tourmaline: showing striations

Tourmaline group

$(Ca,Na,K)(Al,Fe^{2+},Fe^{3+},Li,Mg,Mn)_3(Al,Cr,Fe^{3+},V^{3+})_6$
$Al_6(BO_3)_3Si_6O_{18}(OH,F)_4$

Crystal system Trigonal. **Habit** Crystals usually prismatic, often of curved triangular cross-section. The prism faces are often strongly striated parallel to their length, and the two ends of a crystal are often differently terminated. Parallel or radiating crystal groups are common; also massive. **SG** 3.0–3.2 (increasing with iron content) **Hardness** 7 **Cleavage** Very poor. **Fracture** Conchoidal to uneven. **Color and transparency** Usually black or bluish black; also colorless, blue, pink and green. Some crystals are pink at one end and green at the other. The composition of minerals of the tourmaline group are now referred to the three end members elbaite, $(Na(Li,Al)_3Al_6(BO_3)_3Si_6O_{18}(OH)_4$, liddicoatite, $(Ca(Li,Al)_3Al_6(BO_3)_3Si_6O_{18}(OH)_4)$ and dravite $(Na,Mg_3,Al_6(BO_3)_3Si_6O_{18}(OH)_4)$. Pink varieties are sometimes called rubellite; schorl $(Na,Fe^{2+}_3Al_6(BO_3)_3Si_6O_{18}(OH)_4)$ is black, verdelite green, indicolite blue and dravite is brown: transparent to nearly opaque. **Luster** Vitreous. **Distinguishing features** Prismatic habit, striations, color, triangular cross-section. Care is needed in distinguishing tourmaline from epidote. **Occurrence** Tourmaline commonly occurs in granite pegmatites, or in granites which have undergone metasomatism by boron-bearing fluids. It also occurs in sediments adjacent to such granites, and as an accessory mineral in schists and gneisses.

Hemimorphite (Calamine) $Zn_4Si_2O_7(OH)_2.H_2O$

Crystal system Orthorhombic. **Habit** Crystals tabular; also massive, fibrous, mamillated. **SG** 3.4–3.5 **Hardness** $4\frac{1}{2}$–5 **Cleavage** Prismatic, perfect. **Fracture** Conchoidal to uneven. **Color and transparency** White, sometimes bluish, greenish or brownish: transparent to translucent. **Luster** Vitreous. **Distinguishing features** Crystal form. **Occurrence** Hemimorphite is a secondary mineral found in the oxidized zone of zinc-bearing ore bodies, and in limestones adjacent to such bodies, in association with sphalerite, smithsonite, cerussite, and anglesite. The alternative name, calamine, is also applied to smithsonite.

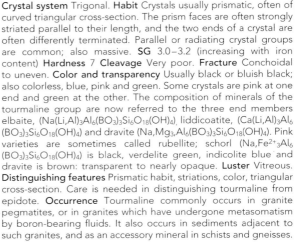

Hemimorphite

2 ins

Tourmaline (rubellite)

Tourmaline

Tourmaline

Hemimorphite

Ilvaite

Hemimorphite

Vesuvianite

2 ins

Vesuvianite (Idocrase) $Ca_{19}Fe(Mg,Al)_8Al_4(SiO_4)_{10}(Si_2O_7)_4(OH)_{10}$
Crystal system Tetragonal. **Habit** Prismatic crystals often with striations parallel to their length; also massive, granular or columnar. **SG** 3.3–3.4 **Hardness** 6–7 **Cleavage** Poor. **Fracture** Subconchoidal to uneven. **Color and transparency** Usually dark green or brown; also yellow. Blue varieties are called cyprine: subtransparent to translucent. **Luster** Vitreous to resinous. **Distinguishing features** Prismatic, striated crystal form. Massive varieties may be mistaken for garnet, epidote or diopside. **Occurrence** Vesuvianite occurs in impure limestones that have undergone contact metamorphism. It occurs in blocks of dolomitic limestone erupted from Vesuvius, hence the name. It is frequently accompanied by grossular, wollastonite, diopside, and calcite.

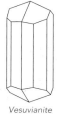

Vesuvianite

Ilvaite $CaFe^{2+}_2Fe^{3+}Si_2O_7O(OH)$
Crystal system Monoclinic, pseudo-orthorhombic. **Habit** Crystals prismatic; also columnar or massive. **SG** 4.1 **Hardness** $5\frac{1}{2}$–6 **Cleavage** One good cleavage. **Fracture** Uneven. **Color and transparency** Black: opaque. **Streak** Black. **Luster** Submetallic **Occurrence** Ilvaite occurs with magnetite in magmatic ore bodies, and in contact metasomatic deposits.

Characteristic cleavages of pyroxenes

Pyroxene group

The pyroxenes are an important and widely distributed group of rock-forming silicates. They have a general formula $X_2Si_2O_6$, in which X is usually Mg, Fe^{2+}, Mn, Li, Ti, Al, Ca or Na. The commonest pyroxenes are Ca, Mg, Fe silicates. The pyroxenes are characterized by two cleavages which intersect almost at right-angles. There are two main groups of pyroxenes: the orthopyroxenes crystallize in the orthorhombic system and contain very little calcium; the clinopyroxenes are monoclinic and contain either Ca, or Na, Al, Fe^{3+} or Li.

Orthopyroxenes:
Enstatite $Mg_2Si_2O_6$ / **Ferrosilite** $(Fe^{2+},Mg)_2Si_2O_6$
Crystal system Orthorhombic. **Habit** Crystals prismatic; usually as grains or massive. **SG** 3.2–4.0 (increasing with iron content) **Hardness** 5–6 **Cleavage** Prismatic, good. **Fracture** Uneven. **Color and transparency** Pale green to dark brownish green, the color usually deepening with iron content. Bronzite and hypersthene are names formerly applied to orthopyroxenes intermediate in composition between enstatite and ferrosilite. Translucent to opaque. **Luster** Vitreous. **Distinguishing features** Two cleavages intersecting almost at right-angles. Pale green color and bronze luster characterizes enstatite and bronzite respectively; hypersthene may resemble clinopyroxene very closely. **Occurrence** Orthopyroxenes are common constituents of igneous rocks such as gabbro and pyroxenite. Orthopyroxenes also occur in some andesitic volcanic rocks and in stony meteorites.

2 ins

Hypersthene

Bronzite

2 ins

Augite

Augite

Diopside

Diopside

Clinopyroxenes:
Diopside-hedenbergite series Ca(Mg,Fe)Si$_2$O$_6$ /
Augite (Ca,Mg,Fe,Ti,Al) (Al,Si)$_2$O$_6$

Crystal system Monoclinic. **Habit** Crystals usually stout prisms of square or eight-sided cross-section; also massive, granular. **Twinning** Common. **SG** 3.2–3.6 (increasing with iron content) **Hardness** $5\frac{1}{2}$–$6\frac{1}{2}$ **Cleavage** Prismatic, good; sometimes a basal parting is present. **Fracture** Uneven. **Color and transparency** Usually dark green to black (augite); diopside is grayish white to light green: translucent to opaque. **Luster** Vitreous. **Distinguishing features** Two cleavages almost at right-angles, crystal form. Diopside is usually a paler green than augite. **Occurrence** Augite is a widely distributed mineral, especially in igneous rocks such as basalt, gabbro, and pyroxenite. Diopside and hedenbergite are characteristic of metamorphic rocks; diopside occurs in metamorphosed impure limestones and skarns and, more rarely, in basaltic igneous rocks; hedenbergite, together with ilvaite and magnetite, commonly occurs in skarns, and in iron-rich sediments which have suffered contact metamorphism.

Augite

Augite: twinned crystal

115

Aegirine

2 ins

Jadeite

Clinopyroxenes: Aegirine $NaFe^{3+}Si_2O_6$

Crystal system Monoclinic. **Habit** Crystals usually slender prisms sometimes terminated by steeply inclined faces giving a pointed appearance; also as discrete grains, or radiating aggregates. **Twinning** Common. **SG** 3.5–3.6 **Hardness** 6 **Cleavage** Prismatic, good. **Fracture** Uneven. **Color and transparency** Dark green or brown often nearly black: subtransparent to opaque. **Luster** Vitreous. **Distinguishing features** Association. **Occurrence** Aegirine occurs typically in sodium-rich igneous rocks such as syenites, nepheline syenites, and associated pegmatites. The name is derived from Aegir, the Scandinavian god of the sea, because the mineral was first described from Norway.

Clinopyroxenes: Jadeite $Na(Al,Fe^{3+})Si_2O_6$

Crystal system Monoclinic. **Habit** Crystals rare, usually massive, granular, compact or columnar. **SG** 3.2–3.4 **Hardness** 6 **Cleavage** Prismatic, good. **Fracture** Splintery. **Color and transparency** Usually various shades of light or dark green; sometimes white: translucent. **Luster** Vitreous; inclined to pearly on cleavage surfaces. **Distinguishing features** Green color, massive habit, tough nature of massive material. The name "jade" is applied to two distinct minerals: jadeite is one, and nephrite, an amphibole, is the other. They are best distinguished by their specific gravity. Softer material, such as serpentine, is often sold as jade. **Occurrence** Jadeite has long been used as a semi-precious and ornamental stone. Jadeite is formed at high pressures: it occurs as grains in metamorphosed sodic sediments and volcanic rocks, and is associated with glaucophane and aragonite.

Clinopyroxenes: Spodumene LiAlSi₂O₆

Crystal system Monoclinic. **Habit** Crystals usually prismatic, often striated along their length, and commonly etched or corroded; also massive, columnar. **Twinning** Common. **SG** 3.0–3.2 **Hardness** $6\frac{1}{2}$–7 **Cleavage** Prismatic, perfect. **Fracture** Uneven, splintery. **Color and transparency** Usually white or grayish white. Hiddenite is a green variety and kunzite is a lilac form; both are used as gemstones. Transparent to translucent. **Luster** Vitreous. **Distinguishing features** Two cleavages. **Alteration** Spodumene is prone to alteration to clay minerals. **Occurrence** Spodumene occurs typically in lithium-bearing granite pegmatites, together with minerals such as lepidolite, tourmaline and beryl. Very large crystals have been recorded, some reaching nearly 50 feet (15 meters) in length and weighing up to 90 tons.

Spodumene (kunzite)

2 ins

Jadeite

Spodumene

Wollastonite $CaSiO_3$

Crystal system Triclinic. **Habit** Crystals tabular or short prismatic; also massive, compact, fibrous or as cleavable masses. **Twinning** Common. **SG** 2.8–3.1 **Hardness** $4\frac{1}{2}$–5 **Cleavage** In three directions, one perfect, two others good. **Fracture** Uneven. **Color and transparency** White to gray: subtransparent to translucent. **Luster** Vitreous; pearly on cleavage surfaces, inclined to silky when fibrous. **Distinguishing features** Color, cleavages, association, dissolves in hydrochloric acid with separation of silica. **Occurrence** Wollastonite occurs in metamorphosed siliceous limestones, either in contact aureoles, in high-grade regionally metamorphosed rocks, or in xenoliths in igneous rocks. It also occurs in certain alkaline igneous rocks. Associated minerals are calcite, epidote, vesuvianite, grossular, and tremolite. It is named after W. H. Wollaston (1766–1828), a British mineralogist.

Rhodonite $(Mn^{2+},Fe^{2+},Mg,Ca)SiO_3$

Crystal system Triclinic. **Habit** Crystals prismatic or tabular but uncommon; usually massive, compact, cleavable or granular. **SG** 3.5–3.7 **Hardness** $5\frac{1}{2}$–$6\frac{1}{2}$ **Cleavage** In three directions, two perfect, one good. **Fracture** Conchoidal to uneven. **Color and transparency** Pink to brown, weathers to black: transparent to translucent. **Luster** Vitreous. **Distinguishing features** Pink color, good cleavages. It resembles rhodochrosite but is harder and unaffected by warm dilute hydrochloric acid. **Occurrence** Rhodonite commonly occurs in association with manganese ore deposits in hydrothermal or metasomatic veins, or in regionally metamorphosed manganese-bearing sediments. It is used as a decorative stone.

Rhodonite

Pectolite $NaCa_2Si_3O_8OH$

Crystal system Triclinic. **Habit** Aggregates of fibrous or acicular crystals, often radiating or stellate. **SG** 2.8–2.9 **Hardness** $4\frac{1}{2}$–5 **Cleavage** Two perfect cleavages. **Fracture** Uneven. **Color and transparency** White: subtranslucent to opaque. **Luster** Silky when fibrous, otherwise vitreous. **Distinguishing features** Acicular form, two cleavages. **Occurrence** Pectolite occurs typically, along with zeolites, in cavities in basalts and similar rocks.

Petalite $LiAlSi_4O_{10}$

Crystal system Monoclinic. **Habit** Crystals rare; usually as masses showing cleavage. **SG** 2.4–2.5 **Hardness** 6–$6\frac{1}{2}$ **Cleavage** One perfect cleavage. **Fracture** Subconchoidal. **Color and transparency** White, gray or green, sometimes colorless or reddish: transparent to translucent. **Luster** Vitreous; pearly on cleavage surface. **Distinguishing features** Petalite resembles cleavage masses of feldspar and is often distinguishable only by optical tests although it gives the red flame characteristic of lithium. The perfect cleavage is sometimes distinctive and the name alludes to this, being derived from the Greek word for a leaf. **Occurrence** Petalite occurs typically in lithium-bearing granite pegmatites along with minerals such as spodumene, tourmaline, lepidolite, and feldspars.

2 ins

Wollastonite

Wollastonite

Rhodonite

Petalite

Pectolite

Characteristic cleavages of amphiboles

Amphibole group

The amphiboles are an important group of rock-forming silicates that are widely distributed in igneous and metamorphic rocks. The angle between the prism faces and between two cleavages parallel to them is about 120°, and is characteristic of the amphiboles. The amphiboles further differ from the pyroxenes in that they are hydrous silicates, the hydroxyl group being an essential part of the structure. The chemical formulae of amphiboles are complex because of the extensive atomic substitution that takes place.

Anthophyllite $(Mg,Fe)_7Si_8O_{22}(OH)_2$ Orthorhombic / **Cummingtonite-grunerite series** $(Fe,Mg)_7Si_8O_{22}(OH)_2$ Monoclinic. **Habit** Individual crystals rare; usually as aggregates of fibrous crystals. **Twinning** Common (cummingtonites). **SG** 2.8–3.4 (anthophyllite): 3.1–3.6 (cummingtonites) (increasing with iron content) **Hardness** 5–6 **Cleavage** Prismatic, perfect. **Color and transparency** White, gray, green, brown. Brownish colors predominate in the cummingtonite series: translucent. **Luster** Vitreous, fibrous varieties silky. **Distinguishing features** Anthophyllite and the cummingtonites are pale in color. Although generally brown, the cummingtonites are so similar to anthophyllite as to require optical or X-ray tests to distinguish them. **Occurrence** Anthophyllite occurs in medium-grade magnesium-rich metamorphic rocks: it does not occur in igneous rocks. Members of the cummingtonite series occur in regionally metamorphosed rocks that are relatively rich in iron and poor in calcium. They occur in contact metamorphic rocks, and cummingtonite occurs in igneous rocks such as certain rhyolites, and as a replacement product of pyroxene in diorites.

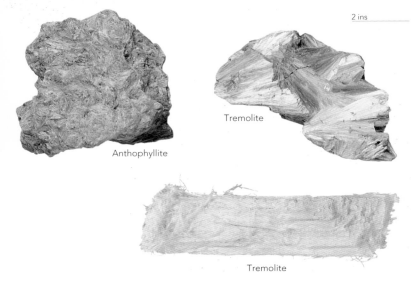

2 ins

Anthophyllite

Tremolite

Tremolite

2 ins

Tremolite

Actinolite

Actinolite

Actinolite

Tremolite-actinolite
(nephrite)

Tremolite-actinolite series $Ca_2(Mg,Fe)_5Si_8O_{22}(OH)_2$
Crystal system Monoclinic. **Habit** Usually in long bladed or prismatic crystals; sometimes massive, fibrous. **Twinning** Common. **SG** 3.0–3.4 (increasing with iron content) **Hardness** 5–6 **Cleavage** Prismatic, good. **Color and transparency** White to gray (tremolite), light to dark green (actinolite), (green color increasing with iron content): transparent to translucent. **Luster** Vitreous. **Distinguishing features** Slender prismatic habit. Fibrous, radiating tremolite resembles wollastonite, but can be distinguished with the microscope and by lack of reaction with hydrochloric acid. Actinolite is paler than most hornblendes. **Occurrence** Tremolite is a characteristic mineral of thermally metamorphosed siliceous dolomitic limestones, and it occurs also in some serpentinites. Actinolite generally occurs in schists produced by low to medium grades of metamorphism of basalt and diabase or of pelitic rocks. It is often fibrous and the name asbestos was originally given to this variety. It also occurs in some igneous rocks, usually as an alteration product of pyroxene. Nephrite, a variety of jade, is usually an actinolitic or tremolitic amphibole.

121

2 ins

Hornblende

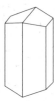

Hornblende

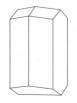

Hornblende: twinned crystal

Hornblende

$Ca_2(Mg,Fe^{2+})_4(Al,Fe^{3+})Si_7AlO_{22}(OH)_2$

Crystal system Monoclinic. **Habit** Crystals usually of long or short prismatic habit; also massive, granular or fibrous. **Twinning** Common. **SG** 3.0–3.5 **Hardness** 5–6 **Cleavage** Prismatic, good. **Fracture** Uneven. **Color and transparency** Light green, through dark green to nearly black; sometimes with a brownish tinge: translucent to nearly opaque. **Luster** Vitreous. **Distinguishing features** The 120° cleavage angle distinguishes hornblende (and other amphiboles) from pyroxene. Hornblende is generally darker than other amphiboles and has a wide range of composition. Hornblende is the name applied to a group of calcic amphiboles whose end-members are magnesiohornblende and ferrohornblende. Chemical analysis is required to distinguish between members of the group. **Occurrence** Hornblende occurs in a wide variety of igneous rocks, being a common constituent of granodiorites, diorites, some syenites and some gabbros, and their fine-grained equivalents. It is also a common constituent of many medium-grade regionally metamorphosed rocks, is particularly characteristic of the amphibolites and hornblende schists, and is commonly accompanied by garnet, quartz and calcic plagioclase.

Glaucophane, riebeckite $Na_2(Mg,Fe^{2+})_3(Al,Fe^{3+})_2Si_8O_{22}(OH)_2$
Crystal system Monoclinic. **Habit** Good crystals rare; often prismatic or acicular; sometimes fibrous. **SG** 3.0−3.4 (increasing with iron content) **Hardness** 5−6 **Cleavage** Prismatic, good. **Fracture** Uneven. **Color and transparency** Glaucophane is gray, gray-blue or lavender-blue; riebeckite is dark blue to black: translucent to subtranslucent. **Luster** Vitreous; fibrous varieties silky. **Distinguishing features** Color and association are distinctive for both glaucophane and riebeckite. **Occurrence** Glaucophane is the magnesian member, poor in ferric iron, of a series with ferroglaucophane. It occurs typically in sodium-rich schists, derived from sediments which have undergone low-temperature/high-pressure regional metamorphism in sub-duction zones. It occurs in association with such minerals as jadeite, aragonite, epidote, chlorite, muscovite and garnet. Riebeckite is the magnesium-poor, ferric iron-rich, member of a series with magnesioriebeckite. It occurs mainly in alkaline igneous rocks such as some granites, syenites and nepheline syenites, and their fine-grained equivalents. Fibrous riebeckite, known as crocidolite or blue asbestos, occurs as veins in bedded ironstones. Riebeckite occurs, though rarely, in schists.

2 ins

Riebeckite (crocidolite)

Riebeckite

Glaucophane

Mica group
There are two main groups of micas: dark mica, rich in iron and magnesium; and white mica which is rich in aluminum. In addition there is a series of lithium micas.

Muscovite $KAl_2(AlSi_3)O_{10}(OH,F)_2$
Crystal system Monoclinic; pseudo-hexagonal. **Habit** Crystals tabular and hexagonal in outline; also as foliated masses and as disseminated flakes. **SG** 2.8–2.9 **Hardness** $2\frac{1}{2}$–3 **Cleavage** Basal, perfect. Cleavage flakes flexible and elastic. **Color and transparency** Colorless to pale gray, green or brown: transparent to translucent. **Luster** Vitreous; pearly parallel to cleavage. **Distinguishing features** Perfect cleavage, light color. **Occurrence** Muscovite is widely distributed. In igneous rocks it is most characteristic of the alkali granites and their pegmatites in which it sometimes forms large masses. It also occurs as a secondary mineral resulting from the decomposition of feldspars. Such fine-grained muscovite is called sericite. It is widely distributed in schists and gneisses and in contact metamorphosed rocks and crystalline limestones. Muscovite survives weathering and transport and is a common constituent of clastic sediments such as sandstones and siltstones.

2 ins

Muscovite

Biotite

Phlogopite

Muscovite

2 ins

Lepidolite

Glauconite

Phlogopite-biotite series:
Phlogopite $K(Mg,Fe^{2+})_3AlSi_3O_{10}(OH,F)_2$ /
Biotite $K(Mg,Fe^{2+})_3(Al,Fe^{3+})Si_3O_{10}(OH,F)_2$
Crystal system Monoclinic. **Habit** Crystals tabular or short pseudo-hexagonal prisms; also as lamellar aggregates or disseminated flakes. **SG** 2.7–3.3 (increasing with iron content) **Hardness** 2–3 **Cleavage** Basal, perfect. **Color and transparency** Phlogopite; yellowish to reddish brown, often with a distinctive coppery appearance; green. Biotite; black, dark brown or greenish black. Transparent to translucent. **Luster** Vitreous; often submetallic on cleavage surface. **Distinguishing features** Perfect basal cleavage; phlogopite is generally paler than biotite. Biotite is the name commonly given to all dark, iron-rich micas. **Occurrence** Phlogopite most commonly occurs in metamorphosed limestones, and in magnesian igneous rocks and some magnesium-rich pegmatites. It occurs also in kimberlite. Biotite is very widely distributed in granite, syenite and diorite and their fine-grained equivalents; it is characteristic of mica lamprophyre. It is a common constituent of schists and gneisses and of contact metamorphic rocks.

Biotite

Glauconite $(K,Na)(Fe^{3+},Al,Mg)_2(Si,Al)_4O_{10}(OH)_2$
Crystal system Monoclinic. Glauconite is a member of the mica family which usually occurs as small, green, rounded aggregates in marine sedimentary rocks. It has a dull luster and a perfect basal cleavage.

Lepidolite $K(Li,Al)_3(Si,Al)_4O_{10}(OH,F)_2$
Crystal system Monoclinic. **Habit** Usually as small disseminated flakes. **SG** 2.8–2.9 **Hardness** $2\frac{1}{2}$–4 **Cleavage** Basal perfect. **Color and transparency** Pale lilac, also colorless, gray or pale pink: transparent to translucent. **Luster** Vitreous; pearly on cleavage surfaces. **Distinguishing features** Perfect cleavage, lilac to pink color. **Occurrence** Lepidolite occurs in granite pegmatites, often in association with lithium-bearing tourmaline and spodumene.

Chlorite

Chlorite group (Mg,Fe,Mn,Al)$_{4-6}$(Si,Al)$_4$O$_{10}$(OH,O)$_8$

Crystal system Monoclinic. **Habit** Crystals tabular pseudo-hexagonal, rarely prismatic; also as scaly aggregates; and massive, earthy. **SG** 2.6–3.3 (increasing with iron content) **Hardness** 2–3 **Cleavage** Basal, perfect. **Cleavage** Flakes are flexible but inelastic. **Color and transparency** Green; also yellow, brown: translucent to subtranslucent. **Luster** Vitreous; earthy in fine-grained masses. **Distinguishing features** Green color and inelastic cleavage fragments distinguish chlorites from micas. The group includes several minerals (e.g. chamosite, clinochlore, pennantite and sudoite) which are difficult to identify without chemical or structural data. **Occurrence** Chlorite occurs in igneous rocks as an alteration product of such minerals as pyroxenes, amphiboles, and micas. It also occurs infilling amygdales in lavas. It is characteristic of low-grade metamorphic rocks and also occurs in sediments.

Serpentine group (Mg,Fe,Ni)$_3$Si$_2$O$_5$(OH)$_4$

Crystal system Monoclinic. **Habit** Serpentine occurs mainly as fibrous chrysotile, the most valued type of asbestos, and as lamellar or platy antigorite. **SG** 2.5–2.6 **Hardness** Variable $2\frac{1}{2}$–4 **Cleavage** Basal, perfect (antigorite); none in fibrous chrysotile. **Fracture** Conchoidal, splintery. **Color and transparency** Various shades of green; also brownish, gray, white or yellow: translucent to opaque. **Luster** Waxy or greasy; fibrous varieties silky; massive varieties earthy. **Distinguishing features** Green color, luster, smooth, rather greasy feel, fibrous habit (antigorite). The serpentine group comprises several polymorphs, the chief of which are antigorite, chrysotile, and lizardite. **Occurrence** Serpentine is a secondary mineral formed from minerals such as olivine and orthopyroxene. It occurs in igneous rocks containing these minerals but typically in serpentinites, which have formed by the alteration of olivine-bearing rocks.

2 ins

Serpentine (chrysotile)

Serpentine (antigorite)

2 ins

Serpentine

Vermiculite

Kaolinite

Vermiculite

Vermiculite $(Mg,Ca)_{0.7}(Mg,Fe^{3+},Al)_6(Al,Si)_8O_{20}(OH)_4.8H_2O$
Crystal system Monoclinic. **Habit** Crystals platy. **SG** About 2.3
Hardness About $1\frac{1}{2}$ **Cleavage** Basal, perfect. **Color and transparency** Yellow, brown: translucent. **Luster** Pearly, sometimes bronzy. **Distinguishing features** Expands greatly perpendicular to cleavage, on heating. **Occurrence** Vermiculite occurs as an alteration product of magnesian micas, in association with carbonatites.

Kaolinite group $Al_2Si_2O_5(OH)_4$
Crystal system Triclinic or monoclinic. **Habit** Microscopic hexagonal plates; usually white earthy masses. **SG** 2.6–2.7 **Hardness** $2–2\frac{1}{2}$ (much less when massive) **Cleavage** Basal, perfect. **Color** White, sometimes grayish or stained brown or red. **Luster** Dull, earthy; crystalline plates pearly. **Distinguishing features** Plastic feel. The kaolinite group includes the polymorphs kaolinite, nacrite, dickite, and halloysite. They cannot be distinguished from other clay minerals without optical or other tests. **Occurrence** Kaolinite is a secondary mineral produced by the alteration of aluminous silicates, and particularly of alkali feldspars.

127

Talc $Mg_3Si_4O_{10}(OH)_2$

Crystal system Monoclinic. **Habit** Crystals rare; usually as granular or foliated masses. **SG** 2.6–2.8 **Hardness** 1 **Cleavage** Basal, perfect. **Color and transparency** White, gray or pale green, often stained reddish: translucent. **Streak** White to very pale green. **Luster** Dull, pearly on cleavage surface. **Distinguishing features** Extreme softness, soapy feel, greenish white color. **Occurrence** Talc occurs as a secondary mineral formed as a result of the alteration of olivine, pyroxene, and amphibole, and it occurs along faults in magnesium-rich rocks. Talc also occurs in schists produced by low- or medium-grade metamorphism of magnesian rocks, often in association with actinolite. Massive talc is called steatite. It occurs rather less frequently as a result of thermal metamorphism of dolomitic limestones.

2 ins

Talc

Apophyllite

2 ins

Prehnite

Apophyllite group KCa$_4$Si$_8$O$_{20}$(OH,F).8H$_2$O

Crystal system Tetragonal. **Habit** Crystals are of varied habit; combinations of prism, bipyramid and pinacoid are most common. **SG** 2.3–2.4 **Hardness** 4½–5 **Cleavage** Basal, perfect; prismatic, poor. **Fracture** Uneven. **Color and transparency** Colorless, white or grayish; sometimes pinkish or yellowish: transparent to translucent. **Luster** Pearly parallel to cleavage; elsewhere vitreous. **Distinguishing features** Crystal form, basal cleavage and pearly luster parallel to it. The basal pinacoid faces are often rough and pitted and contrast with other smooth, bright faces. There is a solid solution series between hydroxyapophyllite and fluorapophyllite. The sodic analogue of the latter is natroapophyllite. **Occurrence** Apophyllite occurs in association with zeolites in cavities in basalt and limestone. It also occurs in some hydrothermal mineral veins.

Apophyllite: combination of prism, bipyramid, and pinacoid

Prehnite Ca$_2$Al$_2$Si$_3$O$_{10}$(OH)$_2$

Crystal system Orthorhombic. **Habit** Crystals rare, tabular; usually in globular and reniform masses with a fibrous structure. **SG** 2.9–3.0 **Hardness** 6–6½ **Cleavage** Basal, good. **Fracture** Uneven. **Color and transparency** Usually pale watery green; also gray, yellow or white: transparent to translucent. **Luster** Vitreous. **Distinguishing features** Green color, habit. **Occurrence** Prehnite occurs most commonly in veins and cavities in igneous rocks, often in association with zeolites. It occurs in very low-grade metamorphic rocks, and as a product of the decomposition of plagioclase feldspar. It is named after Col. H. von Prehn, who discovered the mineral at the Cape of Good Hope, South Africa.

Quartz

Quartz: showing striated prism faces

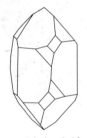

Quartz: right-handed form

Quartz: left-handed form

Silica group

In this group are included those minerals whose composition does not depart significantly from SiO_2. Some varieties are crystalline and include quartz, tridymite, and cristobalite; others, generally grouped as chalcedony, are cryptocrystalline. Opal is amorphous.

Quartz: Japan twin

Quartz: Dauphiné twin

Quartz: Brazil twin

Quartz SiO_2

Crystal system Trigonal. **Habit** Crystals are usually six-sided prisms and are terminated by six faces. The prism faces are often striated at right-angles to the length of the crystal. Imperfectly developed crystals are common. Right- and left-handed forms can be recognized by the presence of small additional faces. **Twinning** Most quartzes are twinned, but twinning is only occasionally observable in crystals. The most common types are Dauphiné twins (double right-handed or double left-handed crystals). Brazil twins (combined right- and left-handed crystals), and Japan twins (contact twins in which the two individuals are nearly at right-angles). **SG** 2.65 **Hardness** 7 **Cleavage** None. **Fracture** Conchoidal. **Color and transparency** Commonly colorless or white, but the range of color is very wide (see below): transparent to translucent. **Luster** Vitreous. **Varieties** Quartz occurs in many varieties. Rock crystal is colorless quartz and sub-varieties include phantom quartz in which growth stages are marked by inclusions, and rutilated quartz (sagenite), which contains hair-like rods of rutile. Amethyst is purple; milky quartz is white; rose quartz is rose-red or pink, and is usually found massive rather than as crystals. Citrine is yellow and transparent and resembles topaz. Smoky quartz (sometimes called cairngorm) is smoky brown to nearly black. Some quartzes contain impurities that not only impart a color but render them opaque. Ferruginous quartz is an example of this and is commonly brick-red or yellow. **Distinguishing features** Crystal form, conchoidal fracture, vitreous luster, hardness. **Occurrence** Quartz is one of the most widely distributed minerals. It occurs in many igneous and metamorphic rocks, particularly in granite and gneiss, and it is abundant in clastic sediments. It is virtually the sole constituent of quartzite. Quartz is also a common gangue mineral in mineral veins, and most good crystals are obtained from this type of occurrence. Well formed quartz crystals can be obtained from cavities (geodes), from granite porphyries, and from granite pegmatites.

2 ins

Quartz
(amethyst)

Rutilated quartz

Quartz (citrine)

Milky quartz

Smoky quartz

Quartz
(rock crystal)

Rose quartz

Sard

Chrysoprase

Chrysoprase

Chalcedony

2 ins

Carnelian

Carnelian

Chalcedony SiO_2

Chalcedony is the name given to compact varieties of silica which comprise minute, often fibrous, quartz crystals with submicroscopic pores. There are two main varieties: chalcedony, which is uniformly colored, and agate which is characterized by curved bands or zones of differing color. **Habit** Often mammillary, botryoidal or stalactitic. Chalcedony commonly has a banded structure, which is not always obvious to the naked eye. It often lines cavities in rocks, and is also massive or nodular. **SG** About 2.6 **Hardness** About $6\frac{1}{2}$ **Cleavage** None. **Fracture** Conchoidal. **Color and transparency** Variable from white through gray, red, brown, to black (*see below*): transparent to subtranslucent. **Luster** Vitreous to waxy. **Distinguishing features** Occurrence and habit, greater density than opal. **Varieties** Various names are applied to the different colored varieties of chalcedony. Carnelian is red to reddish brown and grades into sard which is light to dark brown; chrysoprase is apple-green and heliotrope, also called blood stone, is green with red spots which resemble spots of blood. Jasper is opaque chalcedony and is generally red but yellow, brown, green and gray-blue varieties occur. Jasper is rarely uniformly colored; the color is often distributed in spots or bands. Moss agate consists of a translucent, milky white, bluish white to nearly colorless matrix containing irregularly distributed green, brown, or black moss-like, dendritic impurities of manganese oxide. These often assume attractive figured shapes, and in mocha stone the fern-like branching forms have led to its use in making cameos and other decorative objects. Flint and chert are opaque, granular chalcedony, usually dull gray to black, and which break with a pronounced conchoidal fracture, giving sharp edges. This property was exploited by early man in the fashioning of flint and chert implements. The name chert is used to describe bedded massive chalcedony, and the name flint is reserved for the black nodular variety commonly found in chalk. **Occurrence** Chalcedony is precipitated from silica-bearing solutions and hence forms cavity linings, veins, and replacive masses in a variety of rocks. Chert and flint may originate either by the deposition of silica on the sea floor, or by the replacement of rocks, notably limestone, by silica from percolating waters.

2 ins

Chalcedony

Jasper

Moss agate

Mocha stone

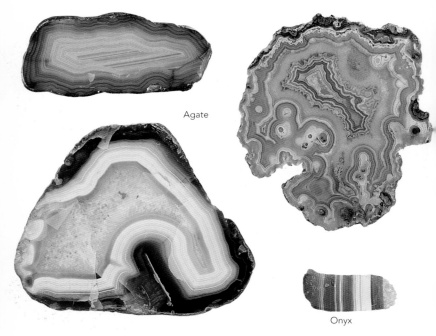

Agate

Onyx

2 ins

Agate SiO₂

Agate is a form of chalcedony which is characterized by bands or zones which differ in color. **Habit** Agate usually forms concentric or irregular layers, usually lining a cavity. **SG** About 2.6 **Hardness** About $6\frac{1}{2}$ **Cleavage** None. **Fracture** Conchoidal. **Color** The bands are usually variegated in shades of white, milky white or gray; also shades of green, brown, red or black. Commercial agate is often colored artificially. Onyx is a form of agate with straight, parallel bands that is used particularly for making cameo brooches. **Occurrence** Agate is widely distributed and occurs typically as a cavity filling in lavas. The layering often follows the form of the cavity and gives place inward to crystals of quartz.

Opal SiO₂.nH₂O

Crystal system None: amorphous. **Habit** Massive; often as stalactitic, botryoidal and rounded forms; also as veinlets. **SG** Variable, 1.8–2.3 **Hardness** $5\frac{1}{2}$–$6\frac{1}{2}$ **Cleavage** None. **Fracture** Conchoidal. **Color and transparency** Variable, from colorless, through milky white, gray, red, brown, blue, green to nearly black. Pale colored forms are common. Transparent to sub-translucent. **Luster** Vitreous to resinous; sometimes pearly. **Distinguishing features** Form, low density. Opal resembles chalcedony in its occurrence, but is less dense and less hard. **Varieties** Opal, though amorphous, has a regular, closely packed,

2 ins

Opal

Wood Opal

three-dimensional arrangement of sub-microscopic spheres of silica. Water (about 6–10%) or air fills the spaces between them. Precious opal has a milky white and sometimes black body color and exhibits a brilliant play of colors usually in blues, reds and yellows. Fire opal is a variety in which red and yellow colors are dominant and produce flame-like reflections when turned. Hyalite is colorless, botryoidal opal; wood opal is wood that has been replaced in part by opaline silica; common opal is the translucent, pale variety that is variously colored but lacks the play of colors of precious opal; and hydrophane is a variety which becomes transparent when immersed in water. Siliceous sinter and geyserite are opaline deposits formed around geysers or by precipitation from hot waters. They generally form stalactitic and delicately filamentous forms of various colors. **Occurrence** Opal is deposited at low temperatures from silica-bearing waters. It can occur as a fissure filling in rocks of any kind, but occurs especially in areas of geysers and hot springs. It can also be formed during the weathering and decomposition of rocks. Opal forms the skeletons of organisms such as sponges, radiolaria, and diatoms. Diatomite, or diatomaceous earth, is a fine-grained sedimentary rock of friable, chalky appearance that is made up in large part of the skeletons of such organisms. The name opal comes from Sanskrit and means "gem" or "precious stone".

*Orthoclase/microcline:
prismatic habit*

*Orthoclase/microcline:
Carlsbad twin*

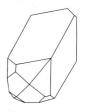

*Orthoclase/microcline:
Baveno twin*

*Orthoclase/microcline:
Manebach twin*

Feldspar group

The feldspars are the most abundant of all minerals in the Earth's crust, and are widely distributed in igneous, metamorphic, and sedimentary rocks. They have the general formula $X(Al,Si)_4O_8$ in which X is Na, K, Ca or Ba. The feldspars may be grouped into the potassic feldspars and the plagioclase feldspars. In the plagioclases, Na and Ca can substitute one for the other. Twinning is common. Carlsbad, Manebach and Baveno twins are simple twins; and albite and pericline twins are repeated twins. In hand specimen, twinning shows as a difference in reflectivity of two halves of a crystal in the case of a simple twin, or as a series of parallel striations of different reflectivity in a repeated twin. Albite twinning is common in the plagioclases. **Alteration** Potassic feldspar alters readily to clay minerals, mainly kaolinite, and the plagioclases usually alter to clay minerals or "sericite".

Potassic feldspars: Sanidine, Orthoclase and Microcline $KAlSi_3O_8$

Crystal system Monoclinic (sanidine and orthoclase); triclinic (microcline). **Habit** Sanidine crystals are usually tabular or prismatic. Orthoclase and microcline are sometimes prismatic, and may be of square cross-section (Baveno habit). **Twinning** Common, on Carlsbad, Baveno or Manebach laws. Microcline also shows repeated twinning on a combination of albite and pericline laws, but this is best seen with a microscope. **SG** 2.5 – 2.6 **Hardness** $6-6\frac{1}{2}$ **Cleavage** Two perfect cleavages. **Fracture** Conchoidal to uneven. **Color and transparency** Sanidine is colorless to gray: transparent. Orthoclase is white to flesh-pink, occasionally red. Microcline is similar to orthoclase, but green varieties are called amazonstone. Both are translucent to subtranslucent. **Luster** Vitreous, rather pearly parallel to cleavage. **Distinguishing features** Orthoclase and microcline are distinguished from other minerals by their color, cleavages and hardness; they are difficult to distinguish one from the other, though green amazonstone is distinctive. Microcline has an ordered Si–Al structure; orthoclase is partially ordered. Sanidine is distinguished by its colorless, transparent appearance, tabular habit and occurrence. Perthite is a potassic feldspar which contains laminae or patches of albite.**Occurrence** Sanidine is the high-temperature, disordered form of $KAlSi_3O_8$ and it occurs as phenocrysts in volcanic rocks such as rhyolite and trachyte. It also occurs in rocks that have been thermally metamorphosed at high temperature. Orthoclase is the common potassic feldspar of most igneous and metamorphic rocks, and microcline, the low-temperature variety, occurs in granites, granite pegmatites, hydrothermal veins, and in many schists and gneisses. Like orthoclase, it is found also as grains in sedimentary rocks.

2 ins

Microcline
(amazonstone)

Sanidine

Orthoclase

Orthoclase

Microcline (perthite)

Microcline on quartz

2 ins

Adularia

Albite

Potassic feldspars: Adularia KAlSi$_3$O$_8$

Crystal system Monoclinic. **Habit** Distinctive single crystals, usually a combination of prism terminated by two faces. **Twinning** Baveno twins common. **SG** 2.6 **Hardness** 6 **Cleavage** Two perfect cleavages. **Fracture** Conchoidal to uneven. **Color and transparency** Colorless or milky white, often with a pearly sheen or play of colors (moonstone): transparent to translucent. **Luster** Vitreous. **Distinguishing features** Simple habit, occurrence. **Occurrence** Adularia is the low-temperature, ordered variety of KAlSi$_3$O$_8$ and occurs in hydrothermal veins.

Plagioclase NaAlSi$_3$O$_8$–CaAl$_2$Si$_2$O$_8$

Crystal system Triclinic. **Habit** Crystals prismatic or tabular; also massive granular. **Twinning** Repeated twinning is common on albite and pericline laws, as are simple twins on Carlsbad, Baveno and Manebach laws. Both simple and repeated twinning may be shown by one individual. **SG** 2.6–2.8 **Hardness** 6–6$\frac{1}{2}$ **Cleavage** Two good cleavages. **Fracture** Uneven. **Color and transparency** Usually white or off-white; sometimes pink, greenish or brownish: transparent to translucent. **Luster** Vitreous, sometimes pearly on cleavage surfaces. **Distinguishing features** Plagioclases are most readily distinguished from potassic feldspars by the presence of repeated albite twin lamellae visible on one of the cleavage

2 ins

Albite

Labradorite

Labradorite

surfaces. The chemistry changes progressively from albite ($NaAlSi_3O_8$) through oligoclase, andesine, labradorite and bytownite, to anorthite ($CaAl_2Si_2O_8$). Individual members of the plagioclase series are difficult to distinguish without a microscope, but labradorite often shows a spectacular play of colors in blues and greens from its cleavage surfaces. For this reason it is often polished and used as a decorative stone. **Occurrence** The plagioclases are widely distributed minerals. They occur in many igneous rocks and are used as a basis of rock classification. In general the sodic plagioclases are characteristic of granitic igneous rocks and give place to more calcic plagioclases in basalts and gabbros. Between potassic feldspar and albite there exists a continuous series as sodium substitutes for potassium; this series is called the alkali feldspar series. In some layered gabbros, plagioclase (usually labradorite or bytownite) forms layers that are virtually free from other minerals, and in places there are large masses of oligoclase-andesine rock called anorthosite. Albite is commonly found in pegmatites and in sodic lavas called spilites. Plagioclase is common in metamorphic rocks, in which it often lacks repeated twinning, and it occurs as detrital grains in sedimentary rocks. Calcic plagioclase occurs in meteorites and in lunar rocks.

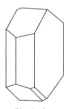

Plagioclase

Plagioclase: repeated albite twinning

139

Feldspathoid group

The feldspathoids are related chemically to the feldspars in that they are sodium and potassium alumino-silicates, but they contain less silica.

Leucite $KAlSi_2O_6$

Crystal system Tetragonal (pseudo-cubic) at ordinary temperatures; cubic above 1157°F. **Habit** Crystals nearly always trapezohedra. **SG** 2.5 **Hardness** $5\frac{1}{2}$–6 **Cleavage** Very poor. **Fracture** Conchoidal. **Color and transparency** Usually white or gray: translucent. **Luster** Vitreous to dull. **Distinguishing features** Crystal form, occurrence. Analcime and garnet also crystallize as icositetrahedra, but analcime occurs typically in cavities, and garnet is neither white nor gray. **Alteration** Leucite may alter to pseudoleucite, a mixture of orthoclase and nepheline. **Occurrence** Leucite does not occur in association with quartz, and is unstable at high pressures, and so it has a restricted occurrence. It occurs typically in potassium-rich, silica-poor lavas such as certain trachytes. Fresh leucite does not occur in plutonic igneous rocks, nor in metamorphic rocks. The name comes from a Greek word meaning "white".

Leucite: trapezohedron

2 ins

Leucite

Leucite

Nepheline

2 ins

Nepheline

Cancrinite

Cancrinite

Nepheline (Na,K)AlSiO₄

Crystal system Hexagonal. **Habit** Crystals usually six-sided prisms; also massive and as discrete grains. **SG** 2.6–2.7 **Hardness** $5\frac{1}{2}$–6 **Cleavage** Prismatic, basal, poor. **Fracture** Conchoidal. **Color and transparency** Usually colorless, white or gray; but also brownish red or greenish: transparent to translucent. **Luster** Greasy to vitreous. **Distinguishing features** Greasy luster, gelatinizes in hydrochloric acid. **Occurrence** Nepheline is characteristic of silica-poor alkaline igneous rocks of both plutonic and volcanic associations. It is found, therefore, in nepheline syenites and ijolites and in lavas such as phonolite. The name comes from the Greek word for a cloud, and alludes to its becoming cloudy when placed in acid.

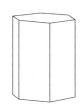

Nepheline

Cancrinite Na₆Ca₂Al₆Si₆O₂₄(CO₃)₂

Crystal system Hexagonal. **Habit** Crystals rare but usually prismatic; generally massive or as discrete grains and as veinlets. **SG** 2.4–2.5 **Hardness** 5–6 **Cleavage** Prismatic, perfect. **Color and transparency** White, gray, yellow, blue: transparent to translucent. **Luster** Vitreous, inclined to greasy. **Distinguishing features** Color, occurrence. **Occurrence** Cancrinite is of restricted occurrence, being found typically in nepheline syenites and associated silica-poor alkaline rocks, in which it generally replaces nepheline. It occurs in some contact metamorphosed limestones. It is named after Count G. Cancrin (1774–1845), a Russian Finance Minister.

Sodalite $Na_8Al_6Si_6O_{24}Cl_2$

Crystal system Cubic. **Habit** Crystals rare, usually rhombdodeca-hedral; commonly massive, granular. **SG** 2.3 **Hardness** $5\frac{1}{2}$–6 **Cleavage** Rhombdodecahedral, poor. **Fracture** Conchoidal to uneven. **Color and transparency** Commonly azure-blue; also pink, yellow, green or gray-white: transparent to translucent. **Luster** Vitreous. **Distinguishing features** Blue color; distinguished from lazurite by its occurrence, and by absence of associated pyrite. Often shows reddish fluorescence in ultra-violet light. **Occurrence** Sodalite occurs with nepheline and cancrinite in alkaline igneous rocks such as nepheline syenites and also in some silica-poor dyke rocks and lavas.

Haüyne $Na_6Ca_2Al_6Si_6O_{24}(SO_4)_2$ Nosean (Noselite) $Na_8Al_6Si_6O_{24}SO_4.H_2O$

Crystal system Cubic. **Habit** Crystals rhombdodecahedra or octahedra; also as discrete grains. **SG** Haüyne 2.4–2.5; nosean 2.3–2.4 **Hardness** $5\frac{1}{2}$–6 **Cleavage** Rhombdodecahedral, poor. **Fracture** Uneven. **Color and transparency** Often blue, also gray, brown, yellow-green: transparent to translucent. **Luster** Vitreous, inclined to greasy. **Distinguishing features** Blue color, association. **Occurrence** Haüyne and nosean both occur typically in silica-poor lavas such as phonolites. Haüyne is named after R. J. Haüy (1743–1822), a French mineralogist. Nosean is named for K. W. Nose (1753–1835), a German mineralogist.

Haüyne/nosean: rhombic dodecahedron

Sodalite

Haüyne

Lazurite

2 ins

Lazurite

Scapolite

Scapolite

2 ins

Lazurite $(Na,Ca)_{7-8}(Al,Si)_{12}(O,S)_{24}(SO_4,Cl,OH)_2$
Crystal system Cubic. **Habit** Crystals rare; cubes or octahedra; commonly massive. **SG** 2.4 **Hardness** $5-5\frac{1}{2}$ **Fracture** Uneven. **Color and transparency** Azure-blue: translucent. **Luster** Vitreous. **Distinguishing features** Color, association with pyrite and calcite. **Occurrence** Lazurite is similar in composition to sodalite, nosean, and haüyne. Lapis-lazuli is a rock rich in lazurite, and is used for jewelry and as a decorative stone. Lapis-lazuli is a contact metamorphosed limestone. Powdered lazurite was once the source of the pigment ultramarine.

Scapolite group $(Na,Ca)_4(Si,Al)_{12}O_{24}(Cl,CO_3,SO_4)$
Crystal system Tetragonal. **Habit** Crystals prismatic often with uneven faces; usually massive, granular. **SG** 2.5–2.8 (increasing with calcium content). **Hardness** 5–6 **Cleavage** Prismatic, in two sets, good; imparts a splintery appearance to massive scapolite. **Fracture** Subconchoidal. **Color and transparency** Usually white or bluish gray; also pink, yellow or brownish: transparent to translucent. **Luster** Vitreous to pearly. **Distinguishing features** Scapolites vary between a sodic end member marialite and a calcic end member meionite. The blocky appearance, pale blue-gray color and splintery, fibrous cleavage are useful features in identification. **Occurrence** Scapolite occurs in metamorphic rocks, particularly in metamorphosed limestones. The mineral also occurs in skarns close to igneous contacts and as a replacement of feldspars in altered igneous rocks.

Scapolite

143

Heulandite

2 ins

Stilbite

Analcime

Zeolites

A group of alumino-silicates containing loosely held water that can be continuously expelled on heating. Members occur as fibrous aggregates while others form robust, non-fibrous crystals.

Analcime (Analcite) $NaAlSi_2O_6.H_2O$

Crystal system Cubic. Habit Crystals usually trapezohedral; also massive. SG 2.2–2.3 Hardness $5\frac{1}{2}$ Cleavage Cubic, very poor. Fracture Subconchoidal. Color and transparency Colorless, white or gray; sometimes tinged with pink or yellow: transparent to subtranslucent. Luster Vitreous. Distinguishing features The crystal form is the same as that of leucite, from which it is distinguished by its occurrence. Occurrence Occurs mainly with other zeolites as a secondary mineral in cavities in basaltic rocks and in sedimentary rocks as a secondary mineral.

Analcime:
trapezohedron

Heulandite $(Na,Ca)_{2\text{-}3}Al_3(Al,Si)_2Si_{13}O_{36}.12H_2O$

Crystal system Monoclinic. Habit Crystals usually tabular and pseudo-orthorhombic. SG 2.1–2.2 Hardness $3\frac{1}{2}$–4 Cleavage One perfect cleavage. Fracture Uneven. Color and transparency White, pink, red or brown: transparent to translucent. Luster Vitreous, pearly parallel to cleavage. Distinguishing features Tabular, coffin-shaped crystals; pearly luster parallel to cleavage, vitreous elsewhere. Occurrence Occurs with stilbite in cavities in basaltic rocks, and in sedimentary rocks as a secondary mineral.

Heulandite

Stilbite $NaCa_2Al_5Si_{13}O_{36}.14H_2O$

Crystal system Monoclinic. Habit Forms sheaf-like aggregates of twinned crystals. Twinning Common, giving cruciform interpenetrant twins. SG 2.1–2.2 Hardness $3\frac{1}{2}$–4 Cleavage One perfect cleavage. Fracture Uneven. Color and transparency White; sometimes yellowish or pink; occasionally brick-red: transparent to translucent. **Luster** Vitreous; pearly on cleavage surfaces. **Distinguishing features** Sheaf-like form, pearly luster parallel to cleavage, vitreous elsewhere. **Occurrence** In cavities in basalts, often with heulandite.

Stilbite: sheaf-like aggregate

Harmotome $(Ba,K)_{1-2}(Si,Al)_8O_{16}.6H_2O$
Crystal system Monoclinic. **Habit** Crystals usually twins, having a pseudo-orthorhombic or pseudo-tetragonal appearance. **Twinning** Very common, interpenetrant. **SG** 2.4–2.5 **Hardness** $4\frac{1}{2}$ **Cleavage** One good cleavage. **Fracture** Uneven. **Color and transparency** White; also yellowish or reddish: subtransparent to translucent. **Luster** Vitreous. **Distinguishing features** Crystal form, occurrence. **Occurrence** In cavities in basalts, often with chabazite.

Chabazite $CaAl_2Si_4O_{12}.6H_2O$
Crystal system Trigonal. **Habit** Crystals rhombohedral but look like cubes. **Twinning** Common; interpenetrant. **SG** 2.0–2.1 **Hardness** $4\frac{1}{2}$ **Cleavage** hedral, poor. **Fracture** Uneven. **Color and transparency** Usually white, yellow; often pinkish or red: transparent to translucent. **Luster** Vitreous. **Distinguishing features** Rhombohedral form. Unlike calcite, does not effervesce with dilute hydrochloric acid. **Occurrence** In cavities in basalts.

Chabazite: rhombohedral habit

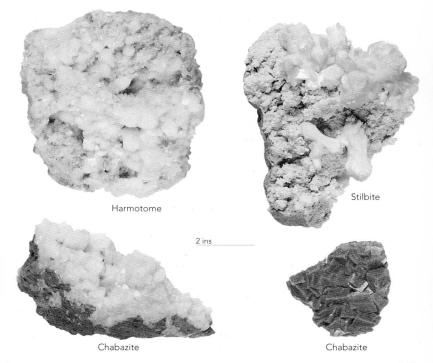

Harmotome

Stilbite

2 ins

Chabazite

Chabazite

Natrolite

Natrolite $Na_2Al_2Si_3O_{10}.2H_2O$

Crystal system Orthorhombic, pseudo-tetragonal. **Habit** Acicular crystals, frequently arranged as divergent or radiating aggregates. **SG** 2.2–2.3 **Hardness** 5 **Fracture** Uneven. **Color and transparency** Colorless or white: transparent to translucent. **Luster** Vitreous. **Distinguishing features** Fibrous habit. **Occurrence** Natrolite occurs typically in the cavities of basaltic and other igneous rocks. Mesolite and scolecite are fibrous zeolites of similar composition and occurrence to natrolite. They are monoclinic but mesolite is pseudo-orthorhombic and scolecite is pseudo-tetragonal. It is difficult to distinguish between them in hand specimen.

Natrolite

2 ins

Mesolite

Thomsonite $NaCa_2Al_5Si_5O_{20}.6H_2O$

Crystal system Orthorhombic, pseudo-tetragonal. **Habit** Acicular crystals in radiating or divergent aggregates. **SG** 2.1–2.4 **Hardness** 5–5½ **Cleavage** Two good cleavages. **Fracture** Uneven. **Color and transparency** White, sometimes tinged with red: transparent to translucent. **Luster** Vitreous to pearly. **Distinguishing features** Similar to natrolite but usually more coarsely crystalline. Difficult to distinguish in hand specimen from other fibrous zeolites. **Occurrence** Thomsonite, like other zeolites, occurs in cavities in lavas, and as a decomposition product of nepheline.

Laumontite $CaAl_2Si_4O_{12}.4H_2O$

Crystal system Monoclinic. **Habit** As small prismatic crystals often with oblique terminations; also massive, or as columnar and radiating aggregates. **SG** 2.2–2.3 **Hardness** 3–3½ **Cleavage** Prismatic, pinacoidal; good. **Fracture** Uneven. **Color and transparency** White; sometimes reddish: transparent to translucent. **Luster** Vitreous; pearly on cleavage surfaces. **Distinguishing features** Laumontite loses part of its water on exposure to dry air and becomes powdery, friable and chalky, when it is known as leonhardite. **Occurrence** Laumontite occurs with other zeolites in veins and amygdales in igneous rocks. It is produced as a result of very low-grade metamorphism of some sedimentary rocks and tuffs. Laumontite is named after G. de Laumont who discovered the mineral.

2 ins

Scolecite

Thomsonite

Laumontite

Rocks

The Earth, and indeed the Moon and planets, are built of the material we call rock. The solid stuff of mountains, the loose sand and gravel of beaches and deserts are all rocks. Rocks are aggregates of minerals, but the petrologist (petrology is the science of rocks), as well as being interested in the mineralogy of rocks, also tries to unravel the record of the geological past which they contain. It is from reading the "record of the rocks" that so much has been learned about past climates and geography, and about the past and present composition of, and the conditions which prevail within, the interior of our planet.

Rocks can be conveniently grouped into *igneous* rocks, *metamorphic* rocks, and *sedimentary* rocks. Igneous rocks are formed by the solidification of molten rock material; metamorphic rocks are formed through the alteration of igneous and sedimentary rocks by heat and pressure; while sedimentary rocks are produced by the accumulation of rock waste at the Earth's surface.

Igneous rocks

The so-called *crust* (Fig. 1) of the Earth is about 22 miles thick under the continents but averages only some 4 miles beneath the oceans. It is formed mainly of rocks of relatively low density. Beneath the crust there is a layer of denser rock called the *mantle* which extends down to a depth of

▲ Fig. 1 Section through the Earth's crust

nearly 1,875 miles. Much of the molten rock material which goes to make up the igneous rocks is generated within the upper parts of the mantle. This material, which is called *magma*, migrates upward into the Earth's crust and forms rock masses which are known as *igneous intrusions*. If magma reaches the Earth's surface and flows out over it, it is called *lava*.

Within some lavas, fragments of dense, green-colored rocks are sometimes found which consist principally of olivine and pyroxene. These fragments (xenoliths) are thought to represent pieces of the mantle, carried upward by the migrating magma.

The great majority of lavas consist of the black, rather dense rock called basalt, and most petrologists consider that the primary molten rock material which comes from the mantle has a composition which is near to that of basalt. Although basalt is the most abundant of the lavas, granite is by far the commonest of the intrusive igneous rocks. Granite is mineralogically and chemically different from basalt and for many years geologists have wrestled with the problem of how the two rock types are related. If basalt is assumed to derive from the mantle, is it likely that granite, which is of a quite different composition, could also come from the mantle? In recent years, ideas as to how basalt, granite and, indeed, the whole spectrum of igneous rocks described in the following pages have been generated, has been much influenced by the theory of *plate tectonics*. In brief, this all-embracing idea is that the uppermost part of the Earth consists of a layer, up to about 94 miles thick, and comprising the crust and the uppermost part of the mantle, that moves over the underlying mantle. This uppermost layer, which is known as the *lithosphere*, comprises a number of rigid "plates" which jostle against each other. In some places new lithosphere is being generated while elsewhere an equivalent amount is being destroyed. New lithosphere is mostly being formed along the *mid-ocean ridges* (spreading ridges), which are lines, notably running along the centre of the Atlantic and within the Pacific Ocean, where magma rises from deep in the mantle, erupts onto the ocean floor and is injected close to the surface, so forming new lithosphere and hence making the plates bigger. Although this activity is generally concealed beneath the waters of the oceans, it can be seen, for instance, in Iceland, which is located on the *Mid-Atlantic Ridge*. Along some plate margins one plate is forced downward beneath the adjacent one. At these boundaries, known as *subduction zones*, the lower plate may be driven hundreds of miles down into the mantle, where the high temperatures will cause melting. This new magma will eventually rise toward the surface where it forms a line of volcanoes. The majority of volcanoes in the world are formed in this way and the many thousands of volcanoes that rim the Pacific Ocean are the result of the edges of the "Pacific Plate" being melted as it is forced down beneath eastern Asia and North and South America.

So, the plate tectonic theory explains where and, in general terms, how igneous rocks are formed. Returning to the problem of basalt and granite, it should be noted that the rocks found along mid-ocean ridges are mainly basalt, while along subduction zones, although basalt is also present, the bulk of the volcanic igneous rocks comprise a spectrum from andesite through to rhyolite, which are chemically equivalent to granite and its associated rocks. Numerous, large bodies of granite also occur in these zones, and represent the intrusive equivalent of the rhyolite. The variety of these igneous rocks occurring above subduction zones reflects the heterogeneity of the melted plates, which are built from most of the rock types found at the surface of the Earth.

Other important places where igneous rocks are produced are "hot spots", which are areas within the plates where heat escaping from the deep Earth is concentrated. In such areas melting may be sufficient to produce volcanoes. Such hot spots form many oceanic islands, for example Hawaii, or volcanoes within the continents, such as Kilimanjaro.

Although the melting of heterogeneous plates explains some of the diversity of igneous rocks found along subduction zones, there is another important process at work. When basalt magma starts to crystallize in the upper mantle, or the lower part of the crust, the overall composition of the crystals is not the same as the overall composition of the magma. This means that the liquid part will have a composition different from that of the original magma, and the further the crystallization process goes the greater will be the difference in composition between the liquid and the crystals. If the crystals and the liquid should now be separated by some mechanism, then rocks of two types will result, each with a composition different from the original basalt. This process, called differentiation, is capable of producing a great range of rock types, one of which is granite.

The recognition and naming of igneous rocks involves an assessment of grain size and the recognition and estimation of the relative amounts of the constituent minerals. Additional information is obtained from color index, texture, structure, and sometimes from field relationships.

Grain size

Grain size refers to the size (average diameter) of the mineral grains comprising the rock. Some rocks have large crystals set in a groundmass of smaller grains (see below); in these rocks only the groundmass minerals are taken into account; the large crystals, no matter how obvious, are ignored. Excluding the glassy rocks, three broad grain size categories are recognized: *fine-grained*, in which the grains are generally below the limit of resolution of the naked eye (less than about 0.004 of an inch); *medium-grained*, in which the grains are recognizable with the naked eye, but minerals hard to identify (0.004 to 0.08 of an inch); and *coarse-grained*, in which the mineral grains can be identified by the naked eye (coarser than 0.08 of an inch). The coarsest rocks, in which the mineral grains have diameters of several inches or more are referred to as *pegmatites*. *Glassy* (or vitric) rocks consist essentially of glass. If magma, or lava, is chilled very rapidly the potential minerals are unable to crystallize and grow, and glass results. The best known such glassy rock is *obsidian*.

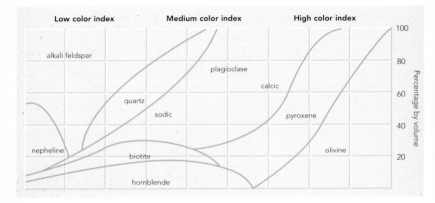

	Low color index			Medium color index			High color index			
nepheline syenite	syenite	granite	grano-diorite	diorite	gabbro	olivine gabbro	peridotite	dunite	Coarse-grained rocks
	micro-syenite	micro-granite	micro-grano-diorite	micro-diorite	diabase (dolerite)	olivine diabase (dolerite)			Medium-grained rocks
phonolite	trachyte	rhyolite	rhyolite	andesite	basalt	olivine basalt			Fine-grained rocks

Mineralogy

This is the most important single feature to be considered when naming igneous rocks. Although magma is a complex silicate melt, most igneous rocks are composed of a few essential minerals belonging to a few mineral groups, namely quartz, the feldspars, and feldspathoids (the light colored or *felsic* minerals), and the pyroxenes, amphiboles, micas, and olivine (the dark colored or *mafic* minerals). Minor constituents are grouped as accessory minerals. All the groups listed are silicates and except for quartz are, within limits, variable in chemical composition. Once the grain size has been decided, rock names are assigned according to the kind and proportions of the constituent minerals. It may be necessary sometimes to know the approximate chemical composition of one particular mineral, but this usually requires at least a microscope and so cannot be determined in the field. The table above summarizes the mineral content and the interrelationships of most of the igneous rocks described in the following pages.

Color index

The color index of a rock is the proportion of dark minerals it contains on the scale 0 to 100.

Texture

Texture refers to the shape, arrangement, and distribution of the minerals of the rock. The following descriptive terms are often used.

In a *granular* texture (equigranular) *(page 158)* all grains are of about the same size and roughly of equant shape. *Poikilitic* texture *(page 166)* refers to large grains of one mineral enclosing smaller grains of other minerals. If pyroxene encloses plagioclase, as in many gabbros and diabases, the texture is called *ophitic*. In a *porphyritic* texture (*page 172*) some large grains (*phenocrysts* or *insets*) are set in a finer grained or glassy matrix (*groundmass*). "Porphyritic" is a common adjectival prefix; for example porphyritic granite, porphyritic basalt.

Flow or *fluidal* texture (*page 170*) refers to tabular or elongate crystals aligned by flow in the magma, in much the same way as logs in a river. In glassy rocks flow is marked by swirling lines, and often by trains of bubbles.

Structure

The structure of rocks refers to the broader features of rock masses rather than those which depend on the interrelationships of the grains.

In a *layered* or *banded* structure (*page 164*) the rock comprises layers of contrasting mineral composition that appear on a surface as bands differing in color or texture. A rock with a *vesicular* structure (*page 177*) contains cavities (*vesicles*) produced by the expansion and escape of gases. Vesicles, which frequently occur in lavas, may be spherical, elliptical, or tubular. When vesicles are filled with secondary minerals they are referred to as *amygdales*, and the structure as *amygdaloidal*. *Xenoliths*, or "inclusions", (*page 163*) are fragments of other rocks included in igneous rocks. They may vary greatly in shape and size. *Joints* are cracks or fissures in rocks along which there has been no displacement. Lava flows sometimes show *columnar jointing* (*page 174*) in which the rock has broken on cooling into parallel hexagonal columns roughly perpendicular to the cooling surface.

Field relationships

Igneous rocks can be divided conveniently into three major groups: *volcanic* (*extrusive*) rocks are largely glassy and fine-grained and form lava flows, tuffs, and agglomerates; *hypabyssal* rocks are largely medium-grained and occur as minor intrusions (sills, dykes); and *plutonic* rocks are largely coarse-grained and form major intrusions. Igneous intrusions are described according to their shapes and their relationships with the rocks they intrude (the *country rocks*) (Fig. 2).

Minor intrusions

Dykes are sheet-like intrusions which are vertical or nearly so and which cut sharply across bedding (*see sedimentary rocks*). Dykes range from a few inches to hundreds of feet in width. *Sills* are sheet-like intrusions which are essentially horizontal and usually follow bedding or foliation. Like dykes they range from a few inches to hundreds of feet in thickness. *Veins* are irregular intrusions which sometimes form a complex network.

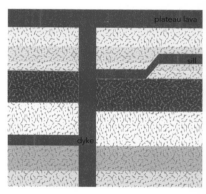

▲ *Fig. 2 Minor igneous intrusions*

Major intrusions

Batholiths (Fig. 3) are large, cross-cutting intrusions, usually of granitic rocks, having steeply dipping contacts and no apparent floor. Exposed batholiths may cover hundreds of thousands of square miles. *Stocks* are smaller than batholiths but otherwise similar. They occupy areas of a few square miles to tens of square miles.

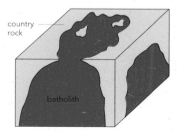

▲ *Fig. 3 Batholith*

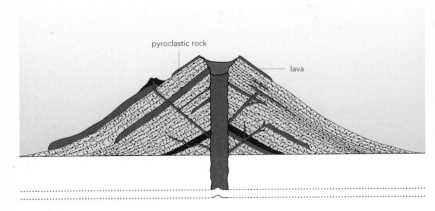

Fig. 4 Section through a volcano

Volcanic rocks

Volcanic cones or volcanoes (Fig. 4) form when lava and accompanying pyroclasts (lava fragments) are ejected from a vertical pipe-like vent. Lava may, however, flow from a *fissure* from which it may travel for considerable distances forming a *lava plateau*. Lava which flows into water chills rapidly and may give rise to distinctive pillow lavas *(page 174)*.

Metamorphic rocks

The Earth's crust as well as being intruded by magma, is from time to time subjected to huge stresses which are sufficiently great to cause it to break to form *faults*, and also to bend forming *folds*. These forces are generally concentrated near to subduction zones, which were described earlier. At these *destructive plate margins* the rocks that form the plates are compressed, folded and heated. The result is a broad, linear zone of rocks which are altered, to varying degrees, by the high temperatures and pressures. This zone of *metamorphic* rocks is generally uplifted to form a mountain chain. The Alps, Himalayas, and Andes, for instance, are all mountain chains that have been generated in this way.

The large scale metamorphism characteristic of many mountain chains is termed *regional metamorphism*. Another type of metamorphism, called *contact metamorphism*, is restricted to the vicinity of igneous intrusions. Magma has a high though variable temperature, usually in the range 1,300 to 2,200°F, so that it heats the rocks adjacent to the intrusion, causing recrystallization and growth of new minerals. The area of altered rocks surrounding an igneous intrusion is called a *contact metamorphic aureole* (Fig. 5), the size of which depends on the temperature of the magma, the size of the intrusion and the nature of the country rocks.

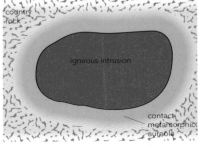

Fig. 5 Contact metamorphic aureole

original rock	Regional metamorphism			Contact metamorphism
	low grade	medium grade	high grade	
quartz sandstone	quartz schist	quartzite	quartzite	quartzite
graywacke	schist	schist	gneiss, granulite	
limestone – pure	marble	marble	marble	marble
limestone – impure	calcareous schist	calc-silicate rock	gneiss	calcareous hornfels
shale/mudstone	slate/phyllite	schist	gneiss granulite	hornfels
diabase/basalt	greenschist	amphibolite	amphibolite, charnockite, eclogite	basic hornfels

Metamorphic rocks display a wide range of texture, structure, and mineralogy, because of the range of temperatures and pressures to which they may have been subjected, and the wide variety of possible rocks from which they were formed (*parental* rocks). The principal types of metamorphic rock and the nature of the parental rocks are summarized in the table above. The terms "low", "medium" and "high" grade refer to the degree of metamorphism that has affected the rocks.

Texture

The minerals of the metamorphic rocks grow in the solid state so that they have to compete for space with the minerals around them, in contrast to minerals of igneous rocks which grow in a fluid (magma). They also often grow under high pressures, and these two factors give to metamorphic rocks their distinctive textures. Most common metamorphic rocks are named accordingly as follows: *slate* is a fine-grained rock with a prominent parting or "cleavage" along which it can be split into thin sheets (*page 182*); *phyllite* is a rock somewhat coarser than slate, but still of fine grain size with a lustrous silvery or greenish sheen on the cleavage surfaces; *schist* is a coarse-grained rock with a marked layering, defined by platy or elongate minerals, often finely interleaved with quartz and feldspar (*page 184*); *gneiss* is a coarse-grained rock

composed largely of quartz and feldspar, but with a marked, though often irregular, layered structure (*page 194*); and *hornfels* is a tough, usually fine-grained, even-textured rock, produced by thermal metamorphism.

Other important textural terms include: *granoblastic*, a texture with mineral grains of the same general size (*page 192*); *porphyroblastic*, a texture in which large, well shaped crystals (*porphyroblasts*) are set in a finer-grained matrix (*page 178*); and *poikiloblastic*, similar to porphyroblastic, but poikiloblasts contain numerous inclusions of another mineral or minerals (*page 179*).

Structure

The structures found in metamorphic rocks are of two types: relict structures inherited from parental sedimentary or igneous rocks which have survived the metamorphism, and new structures produced by the metamorphism itself.

Relict structures inherited from sedimentary rocks include bedding, which may be discernible in a hand specimen, or else as a sequence of different metamorphic rocks that reflect the variation in the original sedimentary succession. Other sedimentary structures such as graded bedding, cross bedding, ripple marks, or even fossils, may be preserved on occasion. Relict structures characteristic of igneous rocks include dykes, fine-grained igneous contacts, pillow structure, and amygdales.

Parental rock	low grade	medium grade	high grade
pelitic rocks (shale/mudstone)	chlorite, biotite	garnet, staurolite	kyanite, sillimanite
impure limestone	calcite, epidote, tremolite	diopside, olivine, grossular garnet	
diabase/basalt	chlorite	garnet, hornblende	

The most significant new metamorphic structures are *cleavage* and *folding (pages 182–183)*. Cleavage is the structure which allows rocks such as slate to be split along parallel planes. It is a product of pressure or *dynamic metamorphism*, and usually cuts across bedding. In schists which are folded on a minor scale cleavages often cut across the bedding in the folds. Heat and high pressure increases the plasticity of rocks so that they tend to yield by folding rather than rupture *(faulting)*. Folds vary from tiny crumples, common in phyllites, and minor folds measured in inches or a few feet, to folds which may be many miles across. The style of the folds is also very variable *(page 185)*.

Mineralogy

Temperature is probably the most important factor in metamorphism and with increasing temperature, even though the rocks are still solid, chemical reactions take place among the rock components and new minerals are produced. In rocks of a particular composition different minerals are produced at different temperatures, so that the mineralogy generally gives a guide to the temperature of formation or *grade of metamorphism*. Contact and regional metamorphic rocks have some significant mineralogical differences, however, probably because of the lower pressures involved in contact metamorphism.

The detailed mineralogical variations among the regionally metamorphosed rocks are rather complex, but a few of the more important minerals, and the metamorphic grades at which they appear, are given in the table above. It must be stressed that most rocks will also contain other minerals, while the "index" minerals will often persist into higher grades.

At the highest grades of regional metamorphism the rocks become plastic and the minerals segregate to produce the banded structure characteristic of gneisses. These rocks are approaching their range of melting, and so grade into the igneous rocks of granitic composition.

The most important minerals occurring in the principal types of contact metamorphosed rocks are set out in the following table.

Parental rock	metamorphic equivalent	minerals – one or more of:
pelitic rocks	hornfels	biotite, andalusite, cordierite, garnet, hornblende, pyroxene, sillimanite, and feldspar
impure limestone	marble, calcareous schist	calcite, dolomite, tremolite, phlogopite, forsterite, diopside, grossular garnet, vesuvianite, and wollastonite
diabase/basalt	basic hornfels	chlorite, hornblende, biotite, garnet, and feldspar

Metasomatism

Although most of the changes in metamorphic rocks take place in the solid state, there is often a movement of material through the rock, carried by migrating fluids. The process by which a rock is changed by the addition and/or subtraction of material through the agency of such fluids is called *metasomatism*, and many metamorphic rocks owe their origin, in part at least, to this process. Good examples are often found among contact metamorphosed limestones, the addition of material to which produces the rocks known as *skarns*, which are often a good hunting ground for the mineral collector.

Sedimentary rocks

Whereas igneous and metamorphic rocks are produced by internal processes within the Earth, sedimentary rocks are formed by processes which are active at the Earth's surface. The surface of the land is continually being attacked by agents of weathering and erosion, such as rain and rivers, wind, and moving ice. These physical agents are helped by chemical decay from percolating waters, and together they break up even the toughest rocks and produce rock waste. This is transported, mainly by rivers but also by wind, and in higher latitudes by ice. Eventually this material, now referred to as sediment, is deposited at river mouths, in lakes, or in the sea. It is the accumulation of this material, often in deposits many miles thick, which goes to make up the sedimentary rocks.

Sedimentary rocks of a different kind are produced from the huge quantities of material carried to lakes and the sea, not as rock or mineral fragments, but dissolved in river water. When the water of lakes or seas becomes saturated with salts, often as a result of evaporation in arid climates, various salts are precipitated and form sedimentary rocks known as *evaporites*.

Yet another group is produced by the gradual accumulation of animal skeletons, such as shells and corals, which are composed essentially of calcium carbonate and go to form the rocks called limestones.

The main features to be considered when studying sedimentary rocks in the field are texture (including grain size), structure, mineralogy, field relationships and, to some extent, color. In the classification used in this guide the sedimentary rocks are first divided into three major divisions according to their

Mechanical origin	
Coarse	conglomerate, breccia, tillite
Medium	sandstone, arkose, orthoquartzite, grit
Fine	siltstone, graywacke, mudstone, shale

Chemical origin	
Calcareous	calcareous mudstone, oolitic and pisolitic limestone (in part), dolostone, travertine
Siliceous	flint, chert
Ferruginous	ironstone
Saline	rock salt, gypsum rock
Phosphatic	phosphate rock (in part)

Organic origin	
Calcareous	biochemical limestone, oolitic and pisolitic limestone (in part)
Carbonaceous	coal
Phosphatic	phosphate rock (in part)

mode of origin: *mechanical origin*, sediments which have been transported as solid particles by water, wind or ice (they are further subdivided according to grain size); *chemical origin*, sediments formed by precipitation from solution of dissolved salts, and sometimes by chemical replacement of one mineral by another (they are further subdivided according to chemical composition); and *organic origin*, sediments formed by the accumulation of organic

155

matter, whether animal or plant (they are further subdivided according to chemical composition).

Texture

This refers to the size, arrangement and shape of the individual grains of the rock. Size categories are: *coarse*, >0.08 of an inch (gravel); *medium*, 0.08 to 0.002 of an inch (sand); and *fine*, within which the range 0.002 to 0.0002 of an inch is called silt; and 0.0002 of an inch, is called clay.

These terms can be applied only loosely in the field, and the distinction between silt (siltstone) and clay (mudstone) is one that requires a microscope, although with practice, it can be made by eye. The shape of individual grains of the medium- and coarse-grained sediments reveals something of their origin. Angular grains, for instance, are unlikely to have been transported very far, whereas rounded grains suggest considerable transport. Sediments with nearly spherical, polished grains are often deposited by wind. The range of grain sizes within a single sediment is also significant. If the range is small, that is, if all grains are about the same size, the sediment is a *mature* one, having been well worked and "sorted" by currents; while a sediment containing a wide range of grain sizes is an *immature* one, and has probably been deposited rapidly or, as in the case of graywacke (*page 199*) and tillite (*page 197*), by a particular mechanism. Other textural features are referred to in the rock descriptions.

Structure

Structure refers to the large scale features which are best seen in the field. These are often particularly informative as to the nature of the environment in which the sediments were accumulated. The most commonly observed structures are described below.

Bedding is of almost universal occurrence in sedimentary rocks, and is a layering expressed by variations of texture and mineralogy. Individual layers were deposited at about the same time, and in an approximately horizontal attitude, although the rocks may have been folded subsequently. Layers having a thickness

greater than 0.5 of an inch are referred to as *strata*, while layers less than 0.5 of an inch thick are *laminae* or *laminations*. The surface separating one stratum or lamination from the next is a *bedding plane*. A collection of strata of one rock type, which can be delimited on a geological map, comprises a *formation*.

Current bedding is a type of bedding in which strata are inclined so that they wedge out at one end and are truncated at the other. The truncated surface is produced by local, contemporaneous erosion (*see unconformity*). Individual current-bedded units may be a few inches or many feet in thickness. Current bedding forms when sediment, commonly sand, is transported by wind or water, and accumulates on a sloping surface, which may be the sea bed at the mouth of a river, or the lee side of a sand-dune. Individual strata wedge out and are inclined in the direction of the current, so that the original direction of flow of wind or water may be determined (*page 199*).

In *graded bedding* the size of grains varies from coarse at the bottom of a stratum to fine at the top. This type of bedding forms by the rapid settling of large quantities of sediment so that larger particles, settling more rapidly, accumulate at the bottom, and the slower settling finer material remains at the top (*see graywacke page 199*). In strongly folded rocks this feature can be used to determine the "right way up".

Slump bedding is a folded or contorted structure produced by the sliding or slumping of wet, recently deposited sediment down a slope on the sea floor (*page 201*). It is often associated with graded bedding.

Ripple marks are commonly seen on bedding planes and are produced by the movement of water or wind over the sediment, thus shaping it into parallel ridges. Cross-sections of individual ripples usually reveal small-scale current bedding (*page 199*).

An *unconformity* in a succession of rocks represents a time interval during which there was erosion, followed later by further deposition. The time gap represented may have been of short duration so that

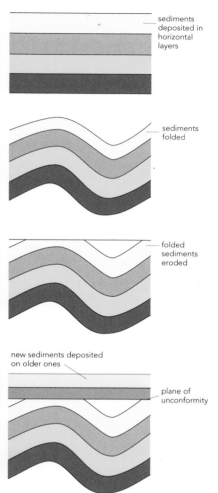

sediments deposited in horizontal layers

sediments folded

folded sediments eroded

new sediments deposited on older ones

plane of unconformity

▲ Fig. 6 Development of an unconformity

distances, and may be of importance in correlating one sequence of rocks with another. They are indicative of relative movements of land and sea (Fig. 6).

For details of organic structures (fossils) see pages 216 to 327, and for details of concretions and nodules see page 208.

Mineralogy

Although there is a wide variation in the chemical composition and mineralogy of sedimentary rocks, these are of less importance in field identification and classification than in the igneous and metamorphic rocks. For help with mineral identification the reader is referred to the mineral section (*pages 6 to 147*).

Field relationships

It is helpful when studying rocks in the field always to examine the relationships between associated rock groups because this will help to elucidate the nature of the environment in which they were formed, and how it changed with time. This in turn may reveal something of the geography of the area at the time of the formation of the sediment (paleogeography).

the older and newer series of rocks are parallel (sometimes called a *disconformity*); sometimes, however, the time span was sufficiently long for there to have been either considerable erosion or for the older rocks to have been tilted before the newer rocks were deposited. Such unconformities can sometimes be traced for considerable

Igneous rocks

Granite

Color Most commonly shades of white, gray, pink and red, but usually mottled in almost any combination of these. **Color index** 0 to 30 **Grain size** Coarse to very coarse; usually constant over large areas. **Texture** Normally granular; commonly porphyritic. Phenocrysts are invariably of feldspar and usually develop good crystal shapes; they often attain sizes of 3 to 4 inches, and they may be aligned owing to flow of the granite magma. Granites may be foliated, due to the parallel arrangement of minerals such as mica and hornblende. **Structure** Granites are typically homogeneous but they may have a banded aspect (*see gneiss on page 194*). Xenoliths are common. Drusy structure is not unusual. Druses are irregular cavities in the rock into which well formed crystals of quartz, feldspar, and other minerals project. Dykes and veins of microgranite, quartz porphyry and pegmatite commonly cut granites. **Mineralogy** Alkali feldspar with or without sodic plagioclase (usually oligoclase) plus quartz, which must comprise 10% or more of the volume of the rock. The feldspar is white or pink; if both a white and a pink feldspar are present, the former tends to be plagioclase and the latter orthoclase. If plagioclase is the dominant feldspar then the rock is granodiorite. Biotite (biotite granite) and/or muscovite (biotite-muscovite or muscovite granite) are usually present, and hornblende may occur. Apatite, titanite, zircon, and magnetite are

Granular texture as seen with a microscope

granite

Cavity with quartz crystals

2 ins

Muscovite granite

Granite with xenolith

common accessories, but a very wide range of other minerals has been recorded from granites. **Field relations** Batholiths, stocks, bosses (a stock of circular cross-section), sills, and dykes. Most of these bodies are clearly intrusive, having sharp, cross-cutting relationships with the country rocks which they metamorphose. In some of the more deeply eroded areas of metamorphic rocks, however, the granites often grade imperceptibly into foliated granite and granite gneiss, and may be partly metamorphic in origin. When sharply cross-cutting, the granite is often finer grained near the contacts owing to more rapid cooling. Irregular veins or *apophyses* commonly extend into the country rocks. Granite often breaks down to various clay minerals (such as kaolin) and quartz sand, but it can also be particularly resistant to breakdown and form rugged outcrops and hills.

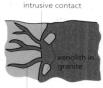

intrusive contact

xenolith in granite

veins (apophyses) in country rock

Contact of granite with country rock

Granodiorite

Color Grays predominate. **Color index** Often a little higher than for granite. **Grain size and texture** As for granite. **Mineralogy** Plagioclase (oligoclase to andesine) more abundant than orthoclase; otherwise as for granite. **Field relations** As for granite. Granodiorites are quantitatively the commonest of the granite family, and indeed, are probably the most voluminous of all the plutonic igneous rocks.

2 ins

Biotite granite

Hornblende-biotite granite

Granite pegmatite

Color White, pink and red, but unevenly colored because of the large size of individual crystals. **Grain size** Very coarse to giant-grained (by definition). Giant crystals have been reported: for example, a spodumene crystal 46 feet long from the Black Hills, South Dakota, USA, and beryl crystals as much as 20 feet long from the same locality. **Texture** The coarseness of grain is striking, but this is usually variable and parts are often finer grained. The crystals often grow parallel or subparallel to each other, and commonly perpendicular to the walls of the intrusion. Graphic or runic texture is common (graphic granite) in which large alkali feldspar crystals contain numerous elongate, regularly spaced, sharply angular quartzes which resemble the characters in certain ancient forms of writing. **Mineralogy** Alkali feldspar and quartz, usually with muscovite. Accessory minerals include all those found in granites and comprise a very varied assemblage. A few of these are beryl, biotite, chalcopyrite, corundum, fluorite, galena, magnetite, oligoclase, allanite, pyrite, pyrrhotite, rutile, titanite, spodumene, topaz, and zircon. **Field relations** Dykes, veins, and irregular segregations, which are usually of rather limited extent. Pegmatites tend to be concentrated in the marginal parts of granite intrusions, or in the country rocks in the immediate vicinity of the granite. Pegmatites are produced by the crystallization of the last residual fluids after the bulk of the granite has solidified, and many of the rarer elements tend to be concentrated in them. They are, therefore, a very important source of many ore minerals for which they are often worked. A wide range of minerals with crystals attaining large sizes occurs in pegmatites, making them the principal single source of the choicest mineral specimens.

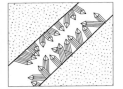

Pegmatite with large crystals growing inward from the walls

2 ins

Tourmaline-bearing pegmatite

Peralkaline granite

Color White, gray and pink. **Color index** 0 to 30 **Grain size**
Coarse. **Texture** Similar to granite but rarely foliated. **Structure**
Similar to granite but xenoliths less common. **Mineralogy** Quartz
(greater than 10%) plus alkali feldspar. Characterized by presence
of alkali-rich pyroxene (aegirine) and/or alkali-rich amphibole
(for example riebeckite). Biotite may also occur. **Field relations**
Stocks, bosses, sills, and dykes. Peralkaline granites are much less
common than granites and granodiorites and do not form such
large intrusions. They tend not to occur in areas of mountain
building, as do the normal granites, but are found in more stable
areas of the continents.

2 ins

Riebeckite granite

Graphic granite

Granite pegmatite

161

Syenite

Color Red, pink, gray or white. **Color index** 0 to 40 **Grain size** Coarse; can be pegmatitic. **Texture** Tends to be equigranular, but also porphyritic and/or fluidal. Essentially similar to granite. **Structure** Drusy cavities common. **Mineralogy** Principally alkali feldspar and/or sodic plagioclase (albite or oligoclase), usually with biotite, amphibole or pyroxene. Up to 10% quartz may occur (quartz syenite), when it grades into granite; or nepheline may be present, when it grades into nepheline syenite. **Field relations** Stocks, dykes, and sills. May be associated with, or grade into, granites, or form individual intrusions which are rarely more than a few miles in diameter. Syenites are not very common rocks.

Syenite

2 ins

Nepheline syenite

Nepheline syenite

Color Usually gray but tones of green, pink, and yellow. **Color index** Usually 0 to 30; but occasionally higher. **Grain size** Coarse; can be pegmatitic. **Texture** Equigranular, porphyritic and/or fluidal. Phenocrysts, when present, are usually of feldspar or nepheline. **Mineralogy** Alkali feldspar and/or sodic plagioclase (albite or oligoclase), nepheline, and often alkali pyroxene or amphibole and/or biotite. Common accessory minerals are cancrinite and sodalite but a great range of rarer minerals has been recorded. **Field relations** Stocks, dykes, and sills. Tend to be associated with other highly alkaline rocks (that is rocks which contain a high proportion of minerals rich in sodium and potassium) such as syenites, and peralkaline granites. Nepheline syenites are relatively rare rocks.

Diorite

Color Speckled black and white in hand specimen; occasionally shades of dark green or pink. The dark minerals are more noticeable than in gabbro. **Color index** 40 to 90, but very variable, often over short distances. **Grain size** Coarse, may be pegmatitic. **Texture** Equigranular or porphyritic. In porphyritic varieties the feldspar or hornblende may form phenocrysts. Diorites often vary rapidly in texture; an equigranular variety may grade into a porphyritic one within a couple of inches. They are sometimes foliated due to the roughly parallel arrangement of the minerals. **Structure** Xenoliths are common. **Mineralogy** Essentially plagioclase (oligoclase or andesine) and hornblende; biotite and/or pyroxene may occur. Alkali feldspar and quartz (quartz diorites) may be present, when diorite grades into granodiorite. Common accessory minerals are apatite, titanite, and iron oxides. **Field relations** Forms independent stocks, bosses, and dykes, but also comprises local variants of masses of granite, and sometimes gabbro, into which they merge imperceptibly.

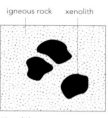

Xenoliths

Diorite

2 ins

2 ins

Olivine gabbro

Ophitic texture shows stippled pyroxene enclosing plagioclase crystals as seen with a microscope. Olivine (heavy outline) and magnetite (black) are also present

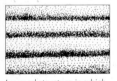

Layered structure in which the tops of the layers are mostly light minerals, and the bottoms of the layers are mostly dark minerals

Layered gabbro intrusion, as seen in cross-section

Gabbro

Color Gray, dark gray, black; may have a bluish or greenish tone. **Color index** 30 to 90; with a decrease of colored minerals gabbro grades into anorthosite, and with an increase it grades into pyroxenite and peridotite. **Grain size** Coarse; can be pegmatitic. **Texture** Granular; porphyritic texture rare. Ophitic texture is common. **Structure** Layering, defined by alternating layers of light and dark colored minerals, often occurs. Individual layers vary from several feet to 0.5 of an inch in thickness. Often contain pegmatitic veins or segregations. **Mineralogy** Essentially plagioclase (labradorite or bytownite) and pyroxene; quartz (quartz gabbro), olivine (olivine gabbro) or hornblende may occur, and iron oxides, chromite, and serpentine are common accessories. **Field relations** Stocks, sills, and dykes. Individual intrusions may be of considerable size (a few miles is usual). Rare, very large sheet-like intrusions, called *lopoliths*, have diameters of hundreds of miles but within these intrusions other rocks such as pyroxenite and anorthosite are also important. Layering is common and field observation indicates that the layers are often arranged like a stack of saucers. If the intrusion has been affected by earth movements such as folding or faulting the layering may be steeply inclined.

Anorthosite

Color Gray to white. **Color index** Less than 10 **Grain size** Medium to coarse. **Texture** Granular. Elongate crystals sometimes occur in parallel alignment and this may be emphasized by streaks and patches of dark minerals. **Structure** May have a layered structure (*see gabbro*). **Mineralogy** Comprises at least 90% plagioclase (oligoclase/andesine to bytownite). Accessory minerals include pyroxene, olivine, and iron oxides. **Field relations** Stocks, dykes, batholiths. In smaller intrusions anorthosites are usually associated with gabbros comprising part of a layered sequence but anorthosites also form huge masses, sometimes covering thousands of square miles, within areas of metamorphic rocks. These masses are variable in composition, grading into gabbro with increase in dark minerals.

Troctolite

Color Gray, studded with black, brown or reddish spots (hence the popular name troutstone). **Color index** 30 to 90 **Grain size** Coarse. **Texture** Granular. **Structure** May have a layered structure, or be part of a layered sequence. **Mineralogy** Plagioclase (labradorite to anorthite) and olivine. The olivine is usually altered to greenish serpentine. **Field relations** Usually associated with gabbro or layered anorthosite.

2 ins

Anorthosite

Troctolite

Layered gabbro

Poikilitic texture showing the small crystals enclosed by larger stippled grain, as seen with a microscope

Peridotite

Color Dull green to black. Dunites (*see below*) are light to dark green or shades of yellow and brown. **Color index** Greater than 90 **Grain size** Medium to coarse. **Texture** Granular. Dunite usually has a sugary texture. Poikilitic texture is common and can be seen by careful inspection of cleavage surfaces. Porphyritic texture is rare. **Structure** Layering may occur. **Mineralogy** Comprises dark minerals only; feldspar is negligble or absent. Olivine is essential (pure olivine rock is called dunite); pyroxene and/or hornblende are usually present. Biotite (mica peridotite), chromite, and garnet sometimes occur. **Field relations** Independent dykes, sills, and small stocks, but also forms parts of large layered gabbroic intrusions together with pyroxenite and anorthosite (*see gabbro*). It is unlikely that a magma of peridotite composition exists but peridotite is thought to be an *accumulate* formed by the crystallization, settling and accumulation of olivine crystals from a gabbro magma. Peridotite occurs also as xenoliths in basalt which may represent fragments brought up from deep layered intrusions, or from the Earth's mantle.

Pyroxenite

Color Green, dark green to black. **Color index** Greater than 90 **Grain size** Medium to coarse. **Texture** Granular. **Structure** May be layered. **Mineralogy** Predominantly clino- or orthopyroxene. Olivine, hornblende, iron oxides, chromite or biotite may be present. Feldspar is minor in amount, or absent. **Field relations** Small independent intrusions – stocks or dykes – and as individual bands in layered gabbro (*see gabbro*).

2 ins

Dunite

Pyroxenite

Serpentinite

Color Grayish green, green to black. Often banded, streaked or blotched in bright greens or reds. **Color index** Greater than 90 **Grain size** Medium to coarse. **Texture** Compact, dull waxy, with a smooth to splintery fracture. **Structure** Often banded; commonly criss-crossed by veins of fibrous chrysotile serpentine. **Mineralogy** Principally serpentine minerals. Olivine, pyroxene, hornblende, mica, garnet, and iron oxides may be present. **Field relations** Stocks, dykes, and lenses. Serpentinites are secondary rocks, having formed by the serpentinization of other rocks, principally peridotite. They commonly occur as pods or lenses in folded metamorphic rocks, probably representing altered olivine-rich intrusions.

Kimberlite

Color Bluish, greenish or black. **Grain size** Coarse. **Texture** Usually porphyritic, with a range of minerals, which tend to be rounded and perhaps broken, forming phenocrysts. **Structure** Xenoliths are common. **Mineralogy** Principally serpentinized olivine, phlogopite, and pyrope garnet; together with orthopyroxene and chrome diopside. Accessory minerals include ilmenite, chromite, and often diamond. Calcite may be abundant. **Field relations** Steeply dipping circular to elliptical pipes, which are rarely more than several hundred feet in diameter, and occasionally as dykes. Kimberlite is one of the only two primary sources of diamonds, the other being the rock called lamproite, which also generally forms pipes.

2 ins

Serpentinite

Kimberlite

Porphyritic microgranite
Color Light to dark gray; yellowish or reddish. **Color index** 0 to 30
Grain size Groundmass medium-grained. **Texture** Porphyritic;
phenocrysts commonly have good crystal shape, and may be
aligned due to flow. **Mineralogy** Essentially the same as granite.
Phenocrysts are quartz and feldspar (white, gray or reddish) with,
more rarely, hornblende or biotite. The groundmass comprises
the same minerals but is generally too fine-grained for them to
be distinguished. **Field relations** Dykes, sills, veins. Commonly
intruded into granite and the surrounding rocks.

2 ins

Porphyritic microgranite

Rhyolite

Pumice

Obsidian

168

Rhyolite

Color Usually light colored; white, gray, greenish, reddish or brownish. The color may be even, or in bands of differing shades. **Grain size** Fine to very fine. **Texture** Frequently shows alternating layers that differ slightly in granularity or color. Phenocrysts not uncommon (porphyritic rhyolite). *Flow banding* is sometimes evident, defined by swirling layers of differing color or granularity, and by aligned phenocrysts. **Structure** Vesicles or amygdales may be present. (Pumice is a highly vesicular variety of rhyolite.) May contain *spherulites* which are spherical bodies, often coalescing, comprising radial aggregates of needles, usually of quartz or feldspar. Spherulites are generally less than 0.2 of an inch in diameter, but they may reach 3 feet or more across. They form by very rapid growth in quickly cooling magma, and by the crystallization of glass. **Mineralogy** As for granite, but rapid cooling results in minute crystals. Phenocrysts of quartz, feldspar, hornblende or mica occur. **Field relations** Flows, dykes, and plugs. Rhyolite (or granite) magma is highly viscous and so flows only very slowly, so that if it is extruded it forms very short, thick flows or is confined as a plug in the throat of a volcano.

Flow banding in rhyolite

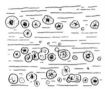

Spherulitic rhyolite

Obsidian and pitchstone

Color Shiny black, also brown or gray. Pitchstone has a dull rather than a shiny luster. **Grain size** None; the rock is glassy. **Texture** Glassy, but obsidian may contain rare phenocrysts; pitchstones contain numerous phenocrysts. **Structure** May be spotted or flow banded and spherulites (*see rhyolite*) are common. Being a siliceous glass, it breaks with a conchoidal fracture and may be fashioned to a sharp cutting edge. It was used for making cutting tools by primitive people. **Mineralogy** Essentially a glass. Rare phenocrysts (abundant in pitchstones) of quartz and feldspar. **Field relations** Dykes and flows. Commonly associated with rhyolites to which they are chemically equivalent.

Obsidian: conchoidal fracture

2 ins

Pitchstone

169

Microsyenite (rhomb porphyry)

Trachyte

2 ins

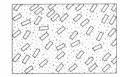

aligned phenocrysts
indicate line of flow

Flow texture

Microsyenite

Color Gray, reddish, pinkish or brownish. **Color index** 0 to 40 **Grain size** Medium. **Texture** Granular; commonly porphyritic. A fluidal texture of closely packed prisms of alkali feldspar is often developed. **Mineralogy** Essentially alkali feldspar; a little biotite, hornblende, pyroxene or quartz is usual. Phenocrysts are usually alkali feldspar; rarely biotite or hornblende. In the type of microsyenite known as rhomb porphyry, the feldspars have characteristic rhomb-shaped cross-sections. **Field relations** Dykes and sills; occasionally lava flows. Microsyenites are associated with intrusions of syenite or nepheline syenite, and with trachyte and phonolite.

Trachyte

Color Usually gray, may be white, pink or yellowish. **Color index** 0 to 40 **Grain size** Fine. **Texture** Almost invariably porphyritic. The rectangular phenocrysts are sanidine. A fluidal texture (known in these rocks as *trachytic*) is characteristically developed, but the fine grain size usually precludes this being seen with the naked eye. **Mineralogy** Dominantly alkali feldspar both in the groundmass and as phenocrysts. A little quartz (less than 10%) or oligoclase may be present. Dark minerals are typically alkaline types such as aegirine or alkali amphibole, and are present only in small amounts so that trachytes are light in color and density. **Field relations** Lava flows and as narrow dykes and sills. Trachyte lavas occur in association with basalts in basaltic volcanoes, but are usually subsidiary to basalt. Occasionally trachyte forms flows of considerable extent.

Phonolite

Leucitophyre

Phonolite

Color Dark green to gray. **Color index** 0 to 30 **Grain size** Fine. **Texture** A rather dense (compact) texture; usually porphyritic. Has a rather greasy luster. **Structure** Often has a platy structure so that it breaks into flat slabs. Reputed to ring when struck with a hammer. **Mineralogy** Alkali feldspar, usually sanidine, nepheline and aegirine or alkali amphibole (such as riebeckite). The phenocrysts are usually rectangular feldspars or nephelines. **Field relations** Lava flows, sills, and dykes. Sometimes associated with trachyte; commonly found in the vicinity of nepheline syenite.

Leucitophyre

Color Gray to dark gray, but may contain white spots in a gray or black groundmass. **Color index** 20 to 70 **Grain size** Medium to fine. **Texture** Invariably porphyritic. The characteristic eight-sided or rounded phenocrysts of leucite are distinctive. **Mineralogy** Alkali feldspar, leucite, and pyroxene; nepheline, phlogopite, and alkali amphibole may be present. **Field relations** Dykes and lava flows. Relatively rare rocks, and associated with other rocks containing feldspathoids.

2 ins

Microdiorite

Color Gray to dark gray; occasionally greenish or pinkish. **Color index** 40 to 90 **Grain size** Medium. **Texture** Usually porphyritic; hence these rocks are sometimes called *porphyrites*. **Mineralogy** As for diorite. Phenocrysts usually hornblende or biotite but may be augite. **Field relations** Dykes and sills, often forming dyke swarms (*see page 174*) in the vicinity of diorite or granite intrusions.

Andesite

Color Shades of gray, purplish, brown, green or almost black. **Grain size** Fine; less commonly partly glassy. **Texture** Frequently porphyritic. **Structure** Flow structures may be evident; can be vesicular or amygdaloidal. **Mineralogy** Only the phenocrysts are recognizable in hand specimen; these are white tabular plagioclase feldspars, plates of biotite mica or prisms of hornblende or augite. The microscope shows the groundmass to consist of plagioclase (oligoclase-andesine) with one or more of the minerals hornblende, biotite, and orthorhombic or monoclinic pyroxene. **Field relations** As lava flows but may also form dykes. Andesite lavas are second in abundance only to basalt. They are usually associated with basalt and rhyolite, and are commonest in areas of mountain building.

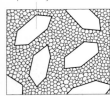

phenocryst

Porphyritic texture

2 ins

Pyroxene lamprophyre

Porphyritic mircodiorite

Andesite

172

Mica lamprophyre

2 ins

Hornblende lamprophyre

Mica lamprophyre

Color Gray to black; often weathers to brownish shades. **Color index** 30 to 70 **Grain size** Medium to fine. **Texture** Porphyritic; rarely granular. Biotite phenocrysts are characteristic and are abundant and of large size, giving the rock a very distinctive appearance. Biotite-rich varieties are noticeably "soft" when hammered. **Mineralogy** The phenocrysts of biotite are readily identified in hand specimen, while more rarely reddish orthoclase and prisms of hornblende may occur. The groundmass, as seen with the microscope, essentially comprises either orthoclase or sodic plagioclase, biotite, and pyroxene or amphibole. Carbonate is often present, when the rock effervesces with dilute hydrochloric acid. **Field relations** Dykes, sills, and small plugs which are usually found in association with granite, syenite or diorite.

Hornblende lamprophyre

Color Greenish, grayish or black when fresh; tends to weather to shades of red or brown. **Color index** 30 to 70 **Grain size** Medium to fine. **Texture** Porphyritic; hornblende phenocrysts usually form long, slender prisms, and are often aligned. Less commonly granular. **Mineralogy** Phenocrysts of hornblende, set in a groundmass consisting of hornblende together with orthoclase or sodic plagioclase. **Field relations** Dykes, sills, and small plugs which occur in the vicinity of granites, syenites, and diorites.

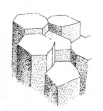

Columnar jointing in basalt

Diabase (Dolerite)

Color When fresh it is black, dark gray or green; may be mottled black and white. **Grain size** Medium. **Texture** Occasionally ophitic texture (*see gabbro*) can be distinguished in hand specimen. May be porphyritic. **Structure** Vesicles and amygdales occur. Sometimes has segregations of coarser rock enriched in feldspar. **Mineralogy** Phenocrysts comprise olivine (olivine diabase) and/or pyroxene or plagioclase. The groundmass comprises the same minerals with iron oxide, and sometimes with some quartz, hornblende or biotite. **Field relations** Dykes and sills. These may form swarms of hundreds or perhaps thousands of individual dykes or sills which often radiate from a single volcanic center.

Basalt

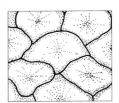

Pillow lava, as seen in cross-section

2 ins

Color When fresh it is black or grayish black; often weathers to a reddish or greenish crust. **Grain size** Fine. **Texture** Usually dense with no minerals identifiable in hand specimen; a freshly broken surface is dull in appearance. May be porphyritic. **Structure** Often vesicular and/or amygdaloidal. Xenoliths are relatively common and usually consist of olivine and pyroxene; they have a green color. Columnar jointing is common and often spectacular, as at the Giant's Causeway, Northern Ireland; individual columns tend to be hexagonal, and up to 2 feet across. *Spheroidal* weathering structures are sometimes developed in which successive layers break away like the skins of an onion leaving a rounded core. **Mineralogy** Phenocrysts are usually olivine (green, glassy), pyroxene (black, shiny) or plagioclase (white-gray, tabular). If olivine is present the rock is called olivine basalt. Microscopic examination shows the groundmass to consist of plagioclase (usually labradorite), pyroxene, olivine, and magnetite, with a wide range of accessory minerals. Amygdales may be filled, or

Diabase

Olivine basalt

Vesicular basalt

partly filled, with zeolites, carbonates or silica, usually in the form of chalcedony or agate. **Field relations** Lava flows and narrow dykes and sills. The edges of dykes or sills are often finer grained than the centres or even glassy, due to rapid cooling on intrusion. Most basalts occur as lava flows either in volcanoes or as extensive sheets building up a *lava plateau*, which may cover hundreds of thousands of square miles, and may be fed by numerous fissures. The surface forms of lavas are of two principal types: smooth or ropy (the surface looks like rope) which is known by the Hawaiian term of *pahoehoe*, and scoriaceous which is rough and clinkery and has the Hawaiian name *aa*. Another common form is *pillow lava* which consists of pillows or balloon-like masses of basalt – usually with a very fine-grained or glassy outer layer. They are formed by the eruption of lava into water.

2 ins

Ropy basalt lava

Amygdaloidal basalt

Pyroclastic rocks

Agglomerate

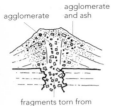

agglomerate and ash

agglomerate

fragments torn from volcanic vent

Section through a volcanic cone

Structure Angular to subrounded fragments, more than 3 inches in diameter, in a finer grained matrix. Most fragments are irregular and highly vesicular; others are rounded, ellipsoidal or spindle-shaped, usually with vesicular interiors, and are called bombs. Bombs are thrown from volcanoes as molten lava clots and acquire their shapes in flight. In modern volcanoes, agglomerate is unconsolidated but becomes consolidated with time. **Composition** Bombs have the composition of the lava erupted by the volcano, such as basalt or andesite, but blocks may also be torn from the sides of the crater or from the rocks beneath the volcano. **Field relations** Agglomerate generally accumulates in the crater of a volcano or on the flanks close to the crater. Associated with tuffs and lavas.

Ash and tuff

Typical shapes of volcanic bombs

Structure Ash comprises unconsolidated volcanic fragments less than 0.08 of an inch in diameter; when consolidated it is called tuff. Ash and tuff commonly contain fragments (up to 3 inches in diameter) which are called lapilli (lapilli ash; lapilli tuff). Lapilli may be angular but are commonly spherical or ellipsoidal having been ejected in a molten state. Ashes and tuffs are usually layered like sedimentary rocks, and often graded within one layer so that larger fragments occur near the base. **Composition** Fragments are essentially of three types: crystalline rock (*lithic* ash, lithic tuff), for example rhyolite, trachyte or andesite; glassy fragments (*vitric*

2 ins

Volcanic bomb

ash, vitric tuff) formed by ejection of liquid lava (these are commonly pumice fragments comprising glass with abundant vesicles); and individual crystals (*crystal* ash, crystal tuff) such as feldspar, augite, and hornblende. Most ashes and tuffs are mixed, including lithic, vitric, and crystal fractions. **Field relations** Ashes are blown from volcanoes during eruptions and with lava flows and agglomerates build up volcanic cones. Generally the larger fragments are found near the crater, while finer grained material may be carried by wind to greater distances, sometimes hundreds of miles from the volcano. They are associated with lavas, agglomerates and sedimentary rocks.

large vesicles (bubbles)

Typical vesicular interior of a volcanic bomb

Ignimbrite

Structure Essentially pieces of pumice, that is highly vesicular glass, usually less than 0.5 of an inch across, in a finer grained matrix of glass fragments. Pumice fragments are often flattened particularly towards the base of flows when they are called *fiamme* (flames). Layering commonly occurs, and columnar jointing is often developed. **Composition** Glass, usually having the composition of rhyolite or trachyte. Some phenocrysts may occur. **Field relations** Ignimbrites are considered to be deposited from ash flows which are ejected rapidly from volcanoes as great volumes of hot expanding gases and incandescent glass fragments. They are controlled by gravity and rush at great speed down the flanks of the volcano.

Ignimbrite: pumice fragments (fiamme)

2 ins

Crystal tuff

Lithic tuff

Ignimbrite

Metamorphic rocks
Contact metamorphism

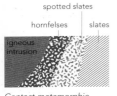

spotted slates

hornfelses | slates

igneous intrusion

Contact metamorphic aureole showing relationship of hornfelses and spotted slates to igneous intrusion

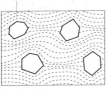

porphyroblast

Porphyroblastic texture

Spotted slate
Color Black, purple, greenish or gray with darker spots. **Texture** Fine-grained, homogeneous; same texture as slate. **Structure** Same structures as slate, that is good cleavage, possibly also sedimentary structures. Characterized by the presence of spots, usually of a spherical or ovoid shape but may be rectangular, up to 0.16 of an inch in diameter. The spots are usually rather vague and filled with inclusions; they may pass gradually into hornfels in which spots are identifiable as cordierite or andalusite. **Mineralogy** Usually too fine-grained for identification in hand specimen, but sometimes the rectangular shapes of andalusite crystals can be identified. **Field relations** In the outer parts of contact metamorphic aureoles involving the metamorphism of pelitic rocks. They usually grade into hornfels of higher grade toward the igneous intrusion.

Andalusite (Chiastolite) cordierite hornfels
Color Black, bluish, grayish, greenish. Often speckled with darker colored porphyroblasts. **Texture** Matrix homogeneous, fine-grained, though a little coarser than slate; contains porphyroblasts or poikiloblasts of andalusite or cordierite which exceptionally may be a couple of inches in diameter. The equigranular texture makes the rock tough and splintery. **Structure** Relict structures from the parental sedimentary rocks are usually obliterated by the metamorphic recrystallization, but occasionally primary bedding may be preserved.

2 ins

Spotted slate

Cordierite hornfels

Mineralogy The matrix is too fine-grained to be distinguished, though tiny flakes of mica may be detected with a lens. Andalusite forms rectangular prisms of square section of a black color, but if of large size, may be reddish. Andalusite is sometimes present as poikiloblasts in which the very fine-grained dark inclusions assume a characteristic cross shape in cross-section, or a thin dark band along the center of longitudinal sections; these crystals are known as chiastolite. Cordierite is rarely as well shaped as andalusite, occurring usually as rounded grains. When well formed, it is hexagonal in section. **Field relations** Contact metamorphic aureoles; grades outward into lower grade rocks such as spotted slate. Andalusite hornfelses may be directly in contact with the igneous intrusion which caused the metamorphism, or hornfelses of higher grade, such as pyroxene or sillimanite hornfels, may be interposed between them.

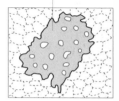

poikiloblast overgrowing and including other minerals

Poikiloblastic texture

Pyroxene hornfels

Color Similar to andalusite-cordierite hornfels. **Texture** Homogeneous, fine- to medium-grained; often with porphyroblasts. **Structure** All primary sedimentary structures destroyed by recrystallization. **Mineralogy** Porphyroblasts of pyroxene, cordierite or andalusite, where present, are usually the only recognizable minerals. **Field relations** The innermost part (that is, highest temperatures) of contact metamorphic aureoles.

Inclusions in andalusite (chiastolite)

Andalusite (chiastolite) hornfels

2 ins

Pyroxene hornfels

Marble

Color White or gray but a wide range of black, red and green occurs, often in streaks and patches. **Texture** Medium- to coarse-grained; granular; often of a sugary appearance. **Structure** Sedimentary structures such as bedding may be preserved. Fossils may be abundant at low metamorphic grades, but with increasing recrystallization they are destroyed. **Mineralogy** Essentially calcite, but may contain greater or lesser amounts of dolomite. Some brucite, olivine, serpentine, tremolite, phlogopite, etc., may be present when it grades into calc-silicate rock and skarn. Marbles are readily scratched with a knife, which serves to distinguish them from the much harder white quartzites. **Field relations** Marbles are produced by the metamorphism of limestone around igneous intrusions. They are thus found in the vicinity of such intrusions but can usually be traced into unmetamorphosed limestone. They are associated with rocks such as hornfelses.

Calc-silicate rock

Color Similar color range to marble but rarely pure white. **Texture** Medium- to coarse-grained; may contain porphyroblasts of calcium silicate minerals, sometimes of large size. **Structure** Original sedimentary bedding may be preserved. **Mineralogy** Calcite together with a wide range of possible minerals including periclase, olivine, serpentine, tremolite, diopside, wollastonite, vesuvianite, and grossular garnet, which are commonly concentrated into bands, patches or nodules. **Field relations** Similar to marble except that calc-silicate rocks are formed by the metamorphism of "impure" limestone containing some sandy or shaly material, which contributes the silicon, aluminum, etc., necessary for the formation of calcium silicate minerals.

Skarn

Color Brown, black, gray but commonly variable even on scale of a hand specimen. **Texture** Fine-, medium- or coarse-grained. **Structure** Minerals often concentrated into layers, nodules, lenses or radiating masses. **Mineralogy** As for calc-silicate rocks, but with iron-rich pyroxene and garnet in addition. Sulfides of iron, zinc, lead, and copper are associated with these silicate minerals. **Field relations** Skarns are produced at the contacts of granite, and sometimes of syenite and diorite, with limestone. Silicon, magnesium and iron from the magma migrates into, and reacts with, the limestone producting the silicate minerals, and sometimes forming ore deposits. Many good mineral specimens are obtained from skarns.

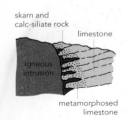

skarn and
calc-siliate rock

limestone

igneous
intrusion

metamorphosed
limestone

Skarn at contact of igneous intrusion, as seen in cross-section

Halleflinta

Color Gray, buff; may be pink, green or brown. **Texture** Fine, even-grained; gives the rock a splintery fracture. **Structure** A layering, after bedding, may be apparent. **Mineralogy** Too fine-grained to distinguish with the naked eye. **Field relations** Represents metamorphosed tuffs which have been impregnated with secondary silica; often found in metamorphic aureoles.

2 ins

Marble

Diopside-garnet marble

Halleflinta

Garnet-diopside skarn

Regional metamorphism

Slate

Color Black and shades of blue, green, brown, and buff. **Texture** Fine-grained. **Structure** By definition, slates are characterized by a single, perfect cleavage (slaty cleavage), enabling it to be split into parallel-sided slabs. On the cleavage surfaces sedimentary structures such as bedding and graded bedding can often be seen. Fossils may be preserved but are invariably distorted. Folds are often apparent in the field. **Mineralogy** Too fine-grained to distinguish with the naked eye. Pyrite porphyroblasts often occur, usually as cubes. **Field relations** Slates are produced by low-grade regional metamorphism of pelitic sediments (shales, mudstones) or fine-grained tuffs. They may be associated with other metamorphosed sedimentary or volcanic rocks.

Phyllite

Color Usually greenish or grayish but with a characteristic silvery sheen. **Texture** Fine- to medium-grained. A well developed schistosity is caused by the parallel arrangement of flaky minerals enabling the rock to be split easily into slabs, and producing the characteristic sheen. **Structure** Minor folds or corrugations are often present. **Mineralogy** Chlorite and/or muscovite are essential constituents, and give to the phyllites their green or gray color. **Field relations** Phyllites are produced from pelitic rocks under conditions of low-grade metamorphism. They usually grade into mica schists.

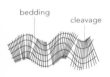

Slate: showing relationship of cleavage to bedding

Slate: showing relationship of cleavage to large-scale folds

2 ins

Phyllite

Slate

Chlorite schist

Color Usually green, may be grayish. **Texture** Fine- to medium-grained. Schistosity well developed. **Structure** Commonly folded on a large or small scale. Larger sedimentary features such as bedding may still be preserved. **Mineralogy** Chlorite, an essential constituent, usually develops as tiny subparallel or parallel flakes indistinguishable to the naked eye, but occasionally forms coarser knots and patches. Porphyroblasts of albite or chloritoid may occur. **Field relations** Chlorite schists form from pelitic rocks under the same grade of metamorphism as phyllites, with which they are associated.

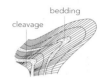

Cleavage developed across a small-scale fold

Glaucophane schist

Color Dark colored with a characteristic purplish or bluish tinge. **Texture** Fine- to medium-grained. Has a schistosity but not usually particularly well developed. Amphibole needles are often in parallel alignment. **Structure** Folding may be apparent. **Mineralogy** Characterized by the presence of glaucophane (a blue alkali amphibole); quartz, albite, jadeite, garnet, and chlorite may also occur. **Field relations** Not common rocks, and probably represent about the same grade of metamorphism as chlorite schists. They are usually formed by high-pressure metamorphism of igneous rocks such as basalt and diabase, but may also be derived from sediments.

2 ins

Chlorite schist

Glaucophane schist

Sericite schist and muscovite schist

Color Usually gray or white, and in coarser varieties the individual mica flakes are brightly reflecting. **Texture** Fine-, medium- or coarse-grained. Invariably has a well developed schistosity. **Structure** Small scale corrugations may occur; sometimes comprises alternate mica-rich and mica-poor layers which may follow the original bedding. **Mineralogy** The dominant mineral is a white mica; if it is very fine-grained it is known as sericite, and the rock is called sericite schist; if the mica is coarser it is referred to by the name muscovite, and the rock is a muscovite schist. The muscovite flakes commonly reach 0.08 to 0.1 of an inch across and are easily prized free with a knife. Quartz is usually present and a little chlorite or garnet may be recognizable. **Field relations** Sericite and muscovite schists represent a moderate grade of metamorphism and tend to be associated with phyllites, chlorite schists, and garnet-bearing schists. They are formed from pelitic sediments, while sandstones containing some argillaceous material give rise to quartz-muscovite schists.

Schistose texture

2 ins

Biotite schist

Quartz-muscovite-biotite schist

Folded biotite schist

Folded sericite schist

2 ins

Biotite schist

Color Usually brownish or black; individual mica flakes may reflect light brightly. **Texture** Coarse-, medium- or fine-grained. A well developed schistosity is invariably present, produced by parallel or subparallel mica flakes. **Structure** May have a banded or striped appearance, caused by variation in the amount of mica present, which can represent original bedding or may be due to *metamorphic segregation*, that is the migration of material within the substance of the rock to form layers of differing chemical and mineral composition. Minor folding common. **Mineralogy** Biotite is the diagnostic mineral of these rocks. The green or brown elastic flakes are easily recognizable, and may be readily prized from the rock with a knife. Biotite is developed partly at the expense of chlorite and/or muscovite, but these minerals may still be present; indeed all proportions of biotite, muscovite, and chlorite may occur. Quartz is usually concentrated into mica-poor layers. Feldspar may be present, commonly in the form of conspicuous white porphyroblasts. **Field relations** Biotite is one of the index minerals of regional metamorphism and its presence indicates a grade of metamorphism higher than that of the chlorite schists. With decreasing metamorphism, biotite schists grade into sericite and chlorite schists and phyllites, while with increasing metamorphism they pass into garnet-bearing schists. Biotite schists are abundant rocks in most regional metamorphic terrains, and represent metamorphosed pelitic sediments.

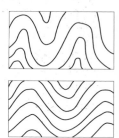

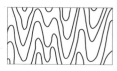

Some different styles of folding

Garnet-mica schist

Color Black, brown, reddish or pinkish. **Texture** Medium- to coarse-grained. A schistosity is invariably well developed. Garnet porphyroblasts are common and may reach half an inch or more across; the schistosity often bends around the porphyroblasts giving an eye-like appearance. **Structure** Folds of various sizes may occur, and a banded appearance due to variations in the amount of mica or concentration of garnet in particular layers is usual. **Mineralogy** Together with the essential garnet, biotite, muscovite, and quartz may be found. The garnet is usually the reddish variety called almandine, and it often forms well shaped crystals that can sometimes be freed from the rock. **Field relations** Garnet-bearing schists are widespread in terrains of regional metamorphism. Garnet is a metamorphic index mineral and indicates a degree of metamorphism higher than that of the biotite schists. Garnet-biotite schists are usually derived from pelitic sediments.

Typical garnet crystals

2 ins

Garnet-biotite schist

Staurolite-biotite schist

Staurolite schist

Color Black, brown, reddish. **Texture** Medium- to coarse-grained. The schistosity is usually good, but may be interrupted by porphyroblasts of staurolite and garnet. The staurolites are often randomly orientated, cutting across the foliation. These rocks are often coarse and banded owing to metamorphic segregation (*see biotite schist*) when they become gneisses. **Structure** May be folded. **Mineralogy** Staurolite forms porphyroblasts with the characteristic prismatic habit and often as twins in the form of a cross. The most commonly associated minerals are garnet, biotite, muscovite, feldspar, and quartz. **Field relations** Staurolite-bearing schists are rocks of comparatively high metamorphic grade, and tend to be associated with other high-grade rocks such as kyanite and sillimanite schists in the central parts of metamorphic belts. They are the product of metamorphism of pelitic sediments.

Albite schist

Color Gray, greenish, brownish. **Texture** Fine-, medium- or coarse-grained. May be phyllitic, schistose or massive. Porphyroblasts of albite are usually conspicuous. **Structure** May be folded. **Mineralogy** White or gray porphyroblasts of albite may be associated with chlorite, epidote, biotite, muscovite or garnet. **Field relations** Rocks containing conspicuous porphyroblasts of albite may be found amongst phyllites and schists of the chlorite, biotite or garnet metamorphic zones. Albite schists probably result from the metamorphism of rocks which originally contained a high proportion of albite, perhaps feldspathic grits or sandstones; or the albite may be a product of metasomatism, having migrated into the schists.

Typical twinned staurolite crystals

2 ins

Garnet-muscovite-biotite schist

Albite schist

Kyanite schist

Color Gray, brown, reddish, with distinctive blue crystals. **Texture** Medium- to coarse-grained. Schistose; may also be gneissose. Kyanite often forms porphyroblasts, sometimes of large size. **Structure** May be folded. **Mineralogy** Kyanite forms sky-blue porphyroblasts of a simple bladed habit which lie parallel to the foliation, or may be concentrated into clusters of crystals; it may occur also in veins with quartz. Other minerals which may be present include garnet, staurolite (either may form porphyroblasts), biotite, muscovite, quartz, and feldspar. **Field relations** Found in the central, high-grade part of metamorphic belts, in association with sillimanite and staurolite schists.

Bladed kyanite crystal

Sillimanite schist

Color Brown, gray, reddish brown. **Texture** Medium- to coarse-grained. Schistosity is not usually striking; may be gneissose. **Structure** May be folded. **Mineralogy** The sillimanite forms needles, which tend to be extremely fine, and slender prisms. Commonly associated minerals include biotite, muscovite, garnet, feldspar, and quartz, and these may be concentrated in contrasting layers. **Field relations** At the highest grades of metamorphism sillimanite replaces kyanite as an index mineral, so that sillimanite schists and gneisses are found in the central parts of metamorphic belts in association with rocks such as kyanite and staurolite schists, and perhaps granites and migmatites.

2 ins

Sillimanite-garnet schist

Garnet-pyroxene granulite

Pyroxene granulite

Color Dark colors – gray, brown. **Texture** Medium- to coarse-grained. Tough, massive rocks which may be layered or banded, but not usually schistose. **Mineralogy** Pyroxene, either hypersthene or diopside, is characteristic. Garnet, kyanite, sillimanite, biotite, hornblende, quartz or feldspar may also be present. **Field relations** The granulites are considered to be formed at very high temperatures and pressures, which imply great depths in the crust. They are, therefore, found in the very old continental shield areas, which have undergone considerable erosion to reveal rocks formed at great depths.

Eclogite

Color Greenish, reddish; sometimes pale to dark green with brownish red spots. **Texture** Medium- to coarse-grained. Massive, banded. Large porphyroblasts of garnet or pyroxene may occur. **Mineralogy** Dominantly composed of pyroxene of a green variety called omphacite, and a reddish garnet. Kyanite and even diamond sometimes occur. The garnet–pyroxene mineralogy is unique and diagnostic. **Field relations** Restricted occurrence being found as lenses and blocks in masses of metamorphic and igneous rocks; particularly in association with peridotites and serpentinites in regions with major faults, and as xenoliths in serpentinite and kimberlite. This suggests that they have been transported from considerable depths, probably from the base of the crust or the mantle. Their very high density – they are among the densest known silicate rocks – supports this hypothesis.

Kyanite-biotite schist

2 ins

Eclogite

Actinolite-chlorite schist

Color Green. **Texture** Fine- to medium-grained; schistosity well developed. **Structure** Primary features of igneous origin such as amygdales may be recognizable. **Mineralogy** Flaky green chlorite and fine green needles of actinolite, which sometimes form radiating clusters or are concentrated into small patches, are characteristic. Feldspar, epidote and calcite may also be identified. **Field relations** Produced by the metamorphism of basic igneous rocks such as diabase and basalt. They occur in phyllites and chlorite schists as cross-cutting layers which trace the position of the original basic dykes, sills or lava flows.

Amphibolite and hornblende schist

Color Black, dark green, green; may be streaked or flecked with white, gray or red. **Texture** Medium- to coarse-grained. A well developed foliation or schistosity may be present, but this is due to stout prisms of hornblende lying in one plane and often aligned within it, rather than to flaky minerals such as micas. The rock does not split as easily as schist. Some varieties are massive, the minerals having no preferred orientation; such rocks have an "igneous" appearance and are conveniently called *epidiorite*. Porphyroblasts, particularly of garnet, may be present. **Structure** A fine banding or layering of alternating dark- and light-colored layers may be developed. These rocks are relatively massive and unyielding so that small-scale folds are rarely developed. **Mineralogy** Amphibole, commonly hornblende but sometimes actinolite or tremolite, comprises some 50% and often as much as 100% of the rock. Actinolite and tremolite tend to form fine needles or long prisms, whereas hornblende forms short, stubby, black shining prisms. Other commonly associated minerals are feldspar (particularly in hornblende schist), chlorite, epidote, pyroxene, and garnet, the latter often forming dark red

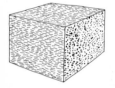

Amphibolite – fabric of subparallel prisms of amphibole

2 ins

Epidiorite

Actinolite-chlorite schist

190

porphyroblasts. A massive rock composed wholly or largely of amphibole is generally called amphibolite, whereas one with a schistosity, and in which there is a large proportion of another mineral, such as feldspar, is called hornblende schist. **Field relations** Result mostly from the metamorphism of igneous rocks such as diabase and basalt, but are of a higher metamorphic grade than actinolite-chlorite schists. As the primary igneous rocks were often sills or dykes cutting sedimentary rocks, so amphibolites and hornblende schists are found as conformable and cross-cutting layers in metamorphosed sediments such as schists, marbles, and quartzites. Amphibolites are tough rocks which tend to resist shearing and so they often occur as disrupted lenses and fragments among schists and gneisses. They are relatively common rocks in most metamorphic terrains.

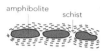

Disrupted layer of amphibolite within schist

2 ins

Garnet-hornblende schist

Hornblende schist

Amphibolite

191

Serpentine marble

Marble

2 ins

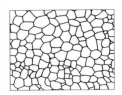

Granoblastic texture

Marble

Color White (the best statuary marble), yellow, red, black or green; either uniform or blotched, banded or veined in differing shades. **Texture** Medium- to coarse-grained; tends to be evenly granular, often sugary in appearance. **Structure** Commonly massive but may have a layering or banding which is usually a primary bedding structure. Schistosity or cleavage is rarely present in pure marbles, but being relatively plastic they may flow readily under high pressures so that folds of a highly contorted type may be developed. **Mineralogy** Calcite and/or dolomite are the major constituents, so that these rocks are relatively soft (they can be scratched easily with a knife which distinguishes them from the harder, white quartzites), and calcite marbles effervesce with dilute hydrochloric acid. If the original limestone contained sand, silt or clay, then the resulting marble contains such minerals as phlogopite, diopside, tremolite, grossular garnet, olivine, serpentine (after olivine) and many others. It is the presence of these minerals that produces the attractive range of colors and structures found in marbles. **Field relations** Marbles are formed by the metamorphism of sedimentary limestones, so they are found in metamorphic terrains in association with other metamorphosed sediments such as quartzite, phyllite and many types of schist. Note that marble can also be formed by contact metamorphism of limestones (*see page 180*).

2 ins

Quartzo-feldspathic schist

Quartzite

Quartzite

Color White, gray, reddish. **Texture** Medium-grained; usually of a granoblastic texture. **Structure** Usually massive but primary sedimentary features may be preserved, such as bedding, graded bedding or current bedding. **Mineralogy** Essentially composed of tightly interlocking grains of quartz. A little feldspar or mica may also be evident. White varieties are distinguished from marble by their greater hardness. **Field relations** Quartzites are metamorphosed quartz sandstones and are found in association with other metamorphosed sedimentary rocks such as phyllite, schist, and marble.

Quartzo-feldspathic schist

Color White, gray, reddish, brownish. **Texture** Medium- to coarse-grained. A poorly developed schistosity is usually apparent, causing the rock to break into slabs. **Structure** May be folded. **Mineralogy** Quartz is a major constituent, together with feldspar. Micas, both muscovite and biotite, usually occur and tend to concentrate in particular layers. **Field relations** Grades, with an increase of quartz, into quartzite, and by a decrease in quartz into schist and phyllite. Represents metamorphosed grit and sandstone which originally contained a high proportion of feldspar, and perhaps also some mica or clay. Occurs in association with other metamorphosed sedimentary rocks such as quartzite, schist, and marble.

193

Gneissose texture

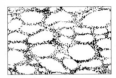

Augen texture

Gneiss and augen gneiss

Color Gray or pink, but with dark streaks and layers. **Texture** Medium- to coarse-grained. Characterized by discontinuous, alternating light and dark layers, the former usually having a coarsely granular texture while the latter, which often contains mica, may be foliated. Augen gneiss contains large porphyroblasts of feldspar, or aggregates of feldspar and quartz, often half an inch or more across, which are eye-shaped, hence the term augen. **Structure** In addition to the gneissose texture described above, gneisses tend to be banded on a large scale with layers and streaks of darker and lighter colored gneiss. Granite and quartz veins and pegmatites are common. May be folded. **Mineralogy** Feldspar is abundant and, together with quartz, forms the granular, lighter colored layers. Muscovite, biotite, and hornblende are commonly present, while any of the minerals characteristic of the higher grades of regional metamorphism may occur. **Field relations** At the highest grades of metamorphism rocks may approach melting temperatures when they are able to recrystallize freely and so produce the textures characteristic of gneisses. Thus gneisses occur, in association with migmatites and granites, in the central parts of metamorphic belts.

2 ins

Folded granite gneiss

Gneiss

2 ins

Granite gneiss with ptygmatic folds

Augen gneiss

Migmatite

Color A mixture of darker colored host rocks and lighter colored (white, pink, gray) granitic rocks. **Texture** Medium- to coarse-grained. The various components of migmatites may display schistosity or gneissose or augen textures. **Structure** Migmatites are mixed rocks comprising a host, usually of schist or gneiss, and a granitic component which may form layers, pods, or veins, or be more evenly distributed through the rock as porphyroblasts of feldspar, or clusters of feldspar and quartz grains (granitization). If the granitization is very extensive, the rock as a whole may approach granite in composition, but original structures, such as layering, folding, etc., are usually still discernible, and are called *phantom structures*. Quartzite, amphibolite, and marble tend to resist granitization, and may form layers and isolated masses in the migmatite. Evidence that these rocks were "plastic" is often provided by swirling folds, and complexly folded granite veins, known as *ptygmatic veins*. **Mineralogy** The host rocks have a mineralogy appropriate to schists or gneisses of high metamorphic grade. The granite fraction comprises essentially alkali feldspar and quartz. **Field relations** Found locally in the inner part of the contact aureoles of large granite intrusions, but on a regional scale in terrains of medium- and high-grade metamorphism, particularly in the Archaean continental shields which, because of the considerable erosion to which they have been subjected, reveal rocks that were formed at considerable depths.

Layering in gneiss

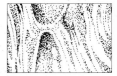

Phantom structure in migmatite

Ptygmatic vein in migmatite

195

Sedimentary rocks

Conglomerate

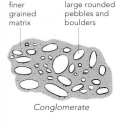

finer grained matrix

large rounded pebbles and boulders

Conglomerate

Color Variable. **Texture** Consists of rounded pebbles (diameter greater than 0.08 of an inch), cobbles or boulders set in a fine- or medium-grained matrix. **Structure** Bedding absent or only crudely developed; may be apparent from variation in the size of the pebbles. Fossils rare. **Mineralogy** Pebbles, boulders, etc., may consist of quartz, chert, flint or almost any igneous, metamorphic or sedimentary rock, but tougher rocks such as quartzite often predominate. The matrix usually comprises sand or silt, often cemented by silica or calcite. **Field relations** Conglomerates are consolidated pebble, gravel or boulder beds which accumulate along sea and lake shores and in rivers. They are indicative of shallow-water sedimentation and vigorous currents, which are required to move large rock fragments. Marine transgressions (rise of sea-level with consequent flooding of the land) are frequently marked by conglomerates which, therefore, are often found immediately above unconformities. Conglomerates are usually associated with sandstone and arkose.

Breccia

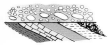

Conglomerate above unconformity

Color Variable. **Texture** Consists of angular rock fragments (one-twelfth of an inch to many feet in diameter) set in a fine- to medium-grained matrix. In some breccias the fragments can be seen to match along their opposed sides, indicating only modest disturbance. **Structure** Bedding not usual, though in some types of breccia, bedding is apparent in the matrix. Fossils

2 ins

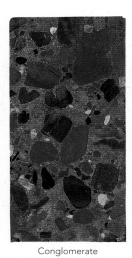

Conglomerate

Conglomerate

rare. **Mineralogy** The fragments may be of any type of igneous, metamorphic or sedimentary rock. The matrix usually consists of silt or sand cemented by calcite or silica. **Field relations** Many breccias represent consolidated talus or scree material, that is, accumulations of rock fragments formerly lying on steep hill slopes, or at the foot of cliffs. They are often found above unconformities, and associated with conglomerate, arkose, and sandstone. Other breccias are produced by the fragmentation of rocks during faulting.

Breccia

Till and tillite
Color Tills are reddish brown or gray; tillites, dark gray to greenish black. **Texture** Angular and some rounded pebbles, cobbles and boulders, characteristically unsorted, that is, of a wide range of sizes, set in a fine- to medium-grained unconsolidated (till) or consolidated (tillite) matrix. Pebbles are often scratched as a result of abrasion during glacial transport. **Structure** Unbedded. **Mineralogy** The rock fragments may be of any type, and in till are in a clayey or sandy matrix, while in tillites the matrix is consolidated into shale or even slate. **Field relations** Till, also known as boulder clay, is deposited by glaciers, and forms widespread surface deposits in the higher latitudes of the northern continents. The recognition of tillites, which are fossil tills, in sedimentary and metamorphic rocks of considerable antiquity (for instance Precambrian, Palaeozoic), indicates that there have been several ice ages.

Ice-scratched pebble

2 ins

Breccia

Tillite

Sandstone, grit, and orthoquartzite

Color Very variable; frequently red, brown, greenish, yellow, gray, white. **Texture** Medium-grained. Usually well sorted, that is, grains all about the same size; grains sharply angular (grit), or subangular to rounded (sandstone). **Structure** Bedding usually apparent; current bedding and ripple marks common; graded bedding may occur. Concretions and fossils may be found. **Mineralogy** Quartz is the main component but is often accompanied by feldspar, mica or other minerals. The grains may be cemented by silica, calcite or iron oxides. Rocks composed almost wholly of quartz grains with a silica cement are known as "pure" sandstones or orthoquartzites. Green glauconite occurs in the variety known as greensand. Sands and sandstones rich in olivine, rutile, magnetite, and other minerals, are found locally. **Field relations** Sandstones are associated with most other sedimentary rocks. Most sands accumulated either in water, usually the sea, or as wind-blown deposits in arid continental areas. Desert sandstones tend to be red, and the individual sand grains are often almost spherical and polished.

angular grains

rounded grains

Sand grains

2 ins

Orthoquartzite

Sandstone

Coarse sandstone (grit)

Arkose

Color Red, pink or gray. **Texture** Medium-grained, but usually nearer to the coarse end of the scale; grains angular. **Structure** Bedding may be obscure or well developed; often current bedded. Fossils rare. **Mineralogy** Contains 25% or more of feldspar, rarely more than 50%; the rest is mainly quartz, but some biotite and muscovite may occur. The cement is usually calcite or iron oxides. **Field relations** Arkoses are derived from the disintegration of granite and granite gneisses, and because they are composed of quartz and felspar they resemble granites, but the angular, fragmental nature of the grains serves to distinguish arkose from the closely interlocking igneous texture of granite. Arkoses occur above unconformities in the immediate vicinity of granite terrains, or in thick deposits associated with conglomerates (containing granite boulders) derived from granites or gneisses.

direction of current

Current bedding

Ripple marking

Graywacke

Color Gray to black, sometimes greenish, usually dark in color. **Texture** Typically contains sharply angular grains (up to 0.08 of an inch) in a finer-grained matrix. **Structure** Frequently massive; graded bedding is typical. Individual graded beds are coarse-grained (sandy, perhaps with a few small pebbles) at the base, and pass upward into silt or clay at the top. Fossils rare. Slump structures common. **Mineralogy** Coarser grains consist of quartz, feldspar, and rock fragments; matrix too fine-grained to be distinguished by the naked eye. The green color is due to the presence of chlorite. **Field relations** Graywackes are deposited rapidly in deep marine basins, by *turbidity currents*; these are masses of sediment-charged water which flow down slopes on the sea bed, and deposit their load of sediment in deep water. Individual units are probably deposited very rapidly, hence their ill-sorted nature.

siltstone or shale (top)

sandstone (bottom)

Graded bedding in graywacke (three units)

2 ins

Arkose

Graywacke

Siltstone

Shale

Siltstone
Color Gray to black, brown, buff, yellow. Texture Grain size 0.002 to 0.0002 of an inch; individual grains can often just be distinguished with the naked eye. Compact, even-textured; but may be earthy. Structure Often shows finely laminated bedding, sometimes picked out by contrasting coloring; may be homogeneous or massive. Fine-scale current bedding and ripple marking also occur. Fossils often abundant, as are nodules and concretions. Mineralogy Too fine-grained for minerals to be distinguished except for rare larger grains of quartz and feldspar. On some bedding surfaces the glint of micas may be seen. Field relations Siltstones form by the compaction of sediment of silt grade which may have accumulated in the sea, in lakes, or be amongst the residual materials of glacial action.

Mudstone, shale, and clay
Color Black, gray, white, brown, red, dark green or blue. Texture Grain size less than 0.0002 of an inch; individual grains

2 ins

Loess

2 ins

Shale

Mudstone

are too small to be distinguished with the naked eye. Mudstone and shale feel smooth, and a pure clay is not gritty when smeared between the fingers. Clays are plastic and often sticky when wet. **Structure** When consolidated and relatively massive it is known as mudstone (or claystone); if finely bedded so that it splits readily into thin layers it is called shale. When soft and uncompacted it is termed clay. Sun cracks, rain prints etc. sometimes occur on bedding surfaces; and fossils and concretions are common. **Mineralogy** Too fine-grained for minerals to be distinguished with the naked eye, or usually even with the microscope. Clays consist of a mixture of clay minerals together with detrital quartz, feldspar, and mica. Iron oxides are usually abundant and contribute to the red and yellow colors. Black shales are rich in carbonaceous matter, and pyrite and gypsum commonly occur in them, sometimes as well shaped crystals. **Field relations** Clays tend only to occur in the younger geological formations, being consolidated into mudstones and shales with time. Being very fine-grained, clay is easily transported by water into the sea and lakes, where it accumulates with silt, sand, and calcareous organisms to form typical sequences of shales, siltstones, sandstones, and limestones. Some clays are *residual*, having formed *in situ* as soils; such are the bauxitic clays (*see mineral section*).

Sun cracks in mudstone

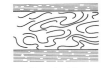

Slump bedding

Aeolian clay (loess)

Color Yellow, brown, buff, gray. **Texture** Fine-grained. Easily powdered in the fingers; compact, earthy, porous. **Structure** Bedding poor. Fossils infrequent. **Mineralogy** Too fine-grained for individual minerals to be seen with the naked eye. **Field relations** Aeolian clays are deposited by wind and the material is probably ultimately of glacial origin. Great thicknesses cover China (loess) and it is widespread elsewhere.

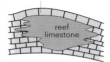

bedded limestone

reef limestone

Reef limestone, as seen in cross-section

Limestone (biochemical)

Color White, gray, cream or yellow when pure; red, brown, black when impure. **Texture** Highly variable from very fine-grained, and porcellaneous, to coarsely crystalline and of sugary appearance. If fossils are present, their abundance and nature partly determine the texture. **Structure** Bedding usually developed. A wide variety of fossils occurs and it is rare not to find some evidence of organic remains. The fossils may be complete, fragmental, or partly destroyed by recrystallization. In richly fossiliferous types the rock usually comprises an assortment of fossil fragments amongst an interstitial finer-grained limestone matrix. Large-scale structures, such as fossil coral reefs with the corals in their original attitudes, may sometimes be observed in large field exposures. Limestones are often criss-crossed by calcite and mineralized veins. **Mineralogy** Essentially comprises finely divided calcite (calcareous mud) or larger crystals which may be derived from animal skeletons such as crinoid plates or by recrystallization, particularly along veins. Finely crystalline silica in the form of chert, and forming bedded or nodular masses, sometimes occurs. Quartz, silt or muddy sediment may be present, and with an increase of these constituents limestones pass into calcareous sandstones and shales, and mudstones. **Field relations** Biochemical limestones are formed principally from the accumulations of the calcareous skeletons of organisms, and they are widely distributed. They form in three principal ways: as reefs,

2 ins

Crinoidal limestone

Chalk

which comprise corals, algal colonies, etc., together with the remains of animals living in and on the reefs such as crinoids and brachiopods; as widespread bedded limestones consisting of the skeletons of bottom-living (benthonic) organisms including many types of gastropods, bivalves, and brachiopods; and as accumulations of the skeletons of floating (pelagic) organisms. The first two types are characteristic of relatively shallow water, while the pelagic limestones may form in deeper water. Some limestones, which can usually be distinguished by the nature of their fossils, are formed in fresh water.

Chalk

Color White, yellow, gray. **Texture** Fine-grained; porous; compact or friable. **Structure** Bedding not usually apparent on small scale. Flint and marcasite nodules common. Fossils usually present. **Mineralogy** Chalk is a very pure limestone formed of calcite, containing only small amounts of silt or mud. Secondary silica (flint) and marcasite are common. **Field relations** Chalk is a pelagic limestone consisting mainly of the tests of coccoliths, foraminifera, and other free-swimming micro-organisms embedded in a fine-grained calcareous mud. It formed in open seas in which there was little or no deposition of other sediments. Most chalks are Cretaceous in age, but similar deposits are accumulating in some parts of the oceans at the present time.

2 ins

Shelly limestone

Fossiliferous freshwater limestone

Oolitic limestone

Pisolitic limestone

Oolitic and pisolitic limestone

Color White, yellow, brown, red. **Texture** Ooliths are spheroidal or ellipsoidal structures built up of concentric layers, and measuring up to 0.08 of an inch in diameter (commonest size about 0.04 of an inch). Larger, more irregular structures, up to pea size, are called pisoliths. Rock composed essentially of closely packed ooliths is called oolite and resembles fish roe. The ooliths may also be dispersed through a finer grained matrix. **Structure** Often current bedded. Fossil fragments occur. **Mineralogy** Usually composed of calcite but ooliths composed of dolomite, silica, and hematite (*see page 206*) also occur. The matrix is also calcite but a few grains of quartz and other detrital minerals are usually present. **Field relations** Ooliths are forming at the present time in certain warm, shallow, strongly agitated parts of the sea, such as the Bahamas Banks. They form by the precipitation and accretion of carbonate around quartz grains and shell fragments rolled along the bottom. Oolitic limestones tend to grade into other limestones and sandstones.

Ooliths in cross-section, as seen with a microscope

2 ins

Calcareous mudstone

Dolostone

Travertine

2 ins

Calcareous mudstone

Color White, gray, yellowish. **Texture** Fine-grained, exhibiting subconchoidal fracture; homogeneous. **Structure** May be bedded; fossils rare. **Mineralogy** Calcite. Some detrital material may be present but it is very fine-grained. **Field relations** Probably formed in relatively deep water, partly as accumulations of the skeletons of free-swimming micro-organisms, and partly as chemical precipitates. The lack of fossils may be due to their recrystallization and breakdown.

Dolostone (Dolomite)

Color White, cream or gray, but often weathers brown or pinkish. **Texture** Coarse, medium or fine; compact, sometimes earthy. **Structure** Bedding tends to be large scale. May be massive or contain complex concretions and nodular growths. Conspicuously jointed. Organic remains usually destroyed by recrystallization. **Mineralogy** Contains a high proportion of dolomite (the name of the mineral and the old name for the rock are the same). Detrital minerals and secondary silica (chert) may be present. **Field relations** Dolostones are usually interbedded with other limestones and are commonly associated with salt and gypsum deposits. Most dolostones are thought to be of secondary (replacement) origin; the calcite of the original limestone having been replaced *in situ* by dolomite, probably by percolating watery solutions.

Travertine and tufa

Color White, yellow, red, brown. **Texture** Compact to earthy; friable. **Structure** Tufa is a porous or spongy rock, while travertine is more dense and often banded. Stalactites are pendant growths from cave roofs, and stalagmites the corresponding floor accumulations; internally they display concentric growth rings. **Mineralogy** Principally calcite; impurities of iron oxides are responsible for yellow and red colors. **Field relations** These rocks are produced by the precipitation of calcite through water evaporation around springs or in caves, where they form thin deposits of no great extent. They are also deposited from water around geysers and hot springs.

Stalactite

Stalagmite

Ironstone
Color Brown, red, green, yellow. **Texture** Fine, medium or coarse; sometimes oolitic. **Structure** Finely and coarsely bedded; also current bedded. Commonly nodular. Organic remains common but usually fragmentary. **Mineralogy** Characterized by the presence of a high proportion of iron-bearing minerals (at least 15% iron) the commonest of which are siderite, hematite, magnetite, pyrite, limonite, chamosite, and glauconite. Detrital minerals usually present, while calcite and dolomite are common cementing agents. **Field relations** They are usually interbedded with cherts, limestones, and sandstones, and may be classified as mudstones, oolites, limestones, sandstones, etc., with suitable mineralogical prefixes. Most ironstones are thought to be chemical deposits, the iron having been precipitated from solution.

2 ins

Chamositic ironstone

Oolitic ironstone

Rock salt

Rock gypsum

Evaporites

Rock salt

Color Colorless, white, orange, red, yellow or rarely purple. **Texture** Massive, coarsely crystalline, glassy or sugary; or as distinct cubic crystals (halite). **Structure** In thick, structureless, massive beds, commonly with partings of shale. Often strongly distorted owing to flow. Fossils rare. **Mineralogy** Essentially halite; easily detected by strong saline taste. Impurities include associated salts (carbonates, sulfates), clay minerals and iron oxides. **Field relations** Formed by the evaporation of saline waters in lagoons, seas, and inland lakes – hence the name evaporites. Particularly associated with shales and dolostones; and with *red beds* (marls and sandstones) which are indicative of formation under desert conditions. Often forms *salt plugs* due to the upward intrusion of the low-density rock salt into overlying sediments.

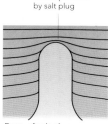
rocks updomed by salt plug

Form of salt plug, as seen in cross-section

Rock gypsum

Color White, pink, red, green or brown. **Texture** Coarse to fine; massive, saccharoidal (sugary) or fibrous; earthy; friable. **Structure** May show bedding, which is often strongly distorted. Usually interbedded with sandstone, mudstone or limestone in which large gypsum crystals occur. Fossils rare. **Mineralogy** Gypsum is commonly associated with anhydrite, halite, calcite, dolomite, clay minerals, and iron oxides. **Field relations** Most gypsum deposits are thought to have been formed by the hydration (addition of water) of anhydrite, which forms in similar environments to rock salt.

Phosphate rock

Color Black, brown, yellow, white. **Texture** Bedded phosphate is fine- to coarse-grained; compact, earthy or granular (sometimes oolitic). Guano is friable; earthy. **Structure** Bedded phosphate rocks are usually nodular and contain organic remains, often replaced by phosphate minerals. Guano is usually bedded. **Mineralogy** Essentially phosphates of calcium, iron, and aluminum, but very complex; associated with common detrital minerals. **Field relations** The most extensive phosphate rock deposits are associated with marine sediments, typically glauconite-bearing sandstones (greensand), limestones, and shales. Guano is an accumulation of the excrement of sea birds and is found principally on oceanic islands.

Phosphate rock

2 ins

Nodules and concretions

Pyrite nodules

Color Bronzy yellow (when freshly broken) but weather to brown, yellow or black. **Texture** When broken open usually reveal radiating, acicular crystals. **Structure** May be spherical, nodular, botryoidal or cylindrical. The surface may be smooth, rough or covered with wart-like knots. **Mineralogy** Pyrite. **Field relations** Pyrite nodules are widespread in a broad range of sedimentary rocks, but particularly in pelitic rocks, especially when these are of a black or blackish gray color. They also occur in limestone.

Flint and chert nodules

Color Blue-gray, gray, to nearly black when fresh, but weather to a whitish, powdery crust (patina). **Texture** Very fine-grained and smooth; conchoidal fracture. Rough on weathered surfaces. **Structure** Flint and chert form rounded nodules of widely differing forms, but chert also forms massive beds. Flint nodules are often hollow and may contain a fossil, such as a sponge or echinoid. **Mineralogy** Composed of silica, mainly the variety chalcedony. Some authors distinguish flint and chert compositionally but the differences, if any, are slight. **Field relations** Flint and chert nodules occur typically in limestone and chalk. They are usually patchily distributed but often concentrated along one bedding plane. Their origin is not fully understood; some appear to be secondary replacements of the host rock, whilst others may represent primary deposition on the seabed of colloidal silica.

cemented concretion

sand grains

Concretion

"Desert Rose" cemented by gypsum

2 ins

Pyrite nodule

Pyrite nodule

Pyrite nodule

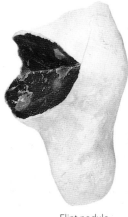

Flint nodule

2 ins

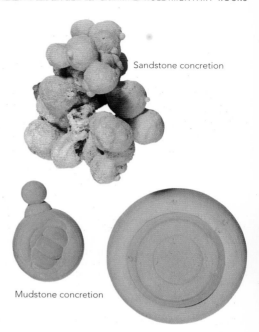

Sandstone concretion

Septarian concretion

Mudstone concretion

Concretions

Color Similar to host rock. **Texture** Similar to host rock. **Structure** Spherical, ellipsoidal, disk-shaped, etc.; with sand crystals the shape is determined by the crystallographic habit of the cementing mineral. Bedding is unbroken from the host rock through the concretion. Concretions vary in size from less than half an inch to several feet across. **Mineralogy** Concretions comprise essentially the same material as the host sediment but are cemented (concreted) together, usually by silica, carbonate or iron oxides. The cement gives the concretion added strength so that it is resistant to weathering and can readily be detached, as a discrete unit, from the surrounding rock. **Field relations** Found in a wide variety of rocks but particularly in shales, siltstones, and sandstones. Concretions form by the local deposition within the sediment, probably by percolating waters, of the cementing mineral. Sand crystals are formed by the crystallization in loose sand of crystals of minerals such as barite, calcite, and gypsum (see desert rose on pages 76–77).

"Desert Rose" cemented by barite

Septaria

Color Black to dark brown or yellow. **Texture** Similar to host sediment. **Structure** Spheroidal to ellipsoidal. Distinguished by a radiating and polygonal pattern of veins, more easily seen when cut open, and which in weathered specimens may stand up as a series of wall-like ridges. **Mineralogy** Pelitic sediment cemented by carbonate; the veins are usually calcite. **Field relations** Found in pelitic sediments. The mechanism of formation is complex and is not fully understood.

carbonate veins

Septaria, as seen in cross-section

209

Meteorites

A meteorite is a natural object that falls to Earth from space and is recovered. Meteorites come in all sizes, from the very tiny, about 0.00004 of an inch across, to the very large, bigger than a house. The smallest ones fall all the time: approximately 40,000 tons of extraterrestrial dust fall on the Earth each year – this is around ten particles per hour per square mile of the Earth's surface. Micrometeorites are small fragments of asteroids and comets that are captured by the Earth. The very smallest of micrometeorites do not melt as they pass through the atmosphere, whereas those that are slightly larger melt and form tiny rounded droplets. Meteors or shooting stars are pieces of dust that burn up high in the atmosphere. Material from a shooting star is not recorded on Earth.

Meteorites fall almost randomly over the Earth's surface, but often are lost – by falling into the ocean, or among other rocks. As few as about ten meteorites that are seen to fall are recovered each year; there are around 2500 authentic meteorites recorded from non-desert areas. Many more meteorites (approximately 15,000 fragments) have been recovered from deserts (both cold and hot), as the dry environment ensures their preservation, and the lack of vegetation and other rocks enhances their chances of being found. More meteorites have been found in Antarctica than anywhere else in the world. This is not because more fall there, but because those that do fall are preserved in the ice, some for as long as up to a million years. The meteorites are carried with the ice as it moves towards the edges of the continent. When the ice meets a barrier that it cannot cross (such as a mountain chain), the ice builds up. Constant scouring of the ice surface by wind reveals concentrations of meteorites in these blue-ice areas. Meteorites have also been found in several hot deserts, for example, the Sahara in Africa and the Nullarbor Plain in Australia. Although they are not concentrated by a specific mechanism, such as ice movement, many meteorites have been recovered from hot deserts, because these are relatively more accessible than the Antarctic. Meteorites that have fallen over the past 50,000 years or so are preserved on the old land surfaces and easily distinguished from the surrounding sand and gravel.

2 ins

Chondrite
showing chondrules

Brecciated achondrite

2 ins

Weathered stony
meteorite

Chondrite
showing ablation crust

Almost all meteorites originate in the Asteroid Belt, which is between the Earth and Mars. However, there are currently 20 or so meteorites that have come from the Moon. Lunar meteorites can be compared directly with samples brought to Earth by the Apollo and Luna missions. The surface of the Moon is covered in craters caused by impacting bodies. The force of these impacts was sometimes sufficient to throw material off the surface of the Moon, and this material arrives on the Earth as a meteorite. In the same way, rocks have come to us from Mars. There are 12 meteorites, the martian origin of which has been inferred on the basis of the composition of gas trapped within pockets of glass inside some of them. The glass was made by melting during shock, probably when the rock was thrown off its parent's surface by an impact. The gas has the same composition as the atmosphere of Mars, as measured by the Viking probes that landed on Mars in 1976, so it is very probable that they are martian in origin.

The importance of meteorites is two-fold: firstly, they provide the best evidence we have for the composition and early history of the solid matter of the Solar System; and secondly, it is con-sidered probable that the composition of some meteorites approaches that of the interior of our planet, about which we have very little direct information.

2 ins

Mineralogy

The commonest minerals that occur in meteorites are of two kinds: the silicates, consisting principally of olivine, pyroxene, and plagioclase; and the nickel-iron alloys, kamacite (Fe, Ni with 4–7% Ni) and taenite (Fe, Ni with 30–60% Ni). The presence of kamacite and taenite is the most outstanding difference between the mineralogy of meteorites and terrestrial rocks, in which they occur only very rarely. The iron sulfide troilite (FeS) is also common in meteorites, but the majority of the 60 or so other minerals recorded are found only in small amounts.

Iron meteorite showing Widmanstätten structure

Weathered
iron meteorite

Classification

Meteorites are classified, principally on their mineralogy, into three groups, namely, irons, stony-irons, and stones. The stones are further subdivided according to the presence or absence of small spherical bodies called *chondrules*. Stones containing chondrules are called chondrites; those without are called achondrites.

Irons

The largest meteorites which have been found are irons, and the biggest of these is estimated to weigh 135,000 lbs. They are usually irregular in shape and may have deep cavities in them, or protuberances from the surface, which may have been caused by collisions in space, fragmentation in flight, weathering, or impact with the Earth's surface. The surface may be smooth, furrowed or covered with shallow depressions. A freshly fallen iron will have a black *fusion crust*, owing to melting by atmospheric friction in flight. This crust is very thin (less than 0.04 of an inch) and is usually confined to one surface. Most irons have a brownish color due to oxidation of the iron during weathering.

The irons consist mainly of nickel-iron alloys. Many irons when cut and etched with acid reveal a complex intergrowth of kamacite and taenite known as *Widmanstätten structure*, which is found only in meteorites.

Stony-irons

The meteorites of this group are composed of nickel-iron and silicate minerals in about equal proportions, and usually consist either of well shaped crystals of olivine in a continuous matrix of nickel-iron, or of plagioclase and pyroxene set in a discontinuous nickel-iron matrix. The stony-irons comprise only about 4% of the known meteorites. The surface features and weathering described for the irons apply also to the stony-irons, but in the latter, particularly the olivine-bearing ones, the silicate minerals may weather out preferentially giving the surface a rough, pitted aspect.

Stones

About 90% of all meteorites which have been observed to fall are stones; more than 90% or more of these are chondrites. Most chondrites have a composition of about 30% pyroxene, 40% olivine, 10% plagioclase, 5–20% nickel-iron, and 6% troilite. The chondrules are composed mainly of olivine or orthopyroxene and may be abundant or sparse. The achondrites are coarser grained than the chondrites, and resemble some terrestrial rocks in texture and mineralogy. They are rather variable in composition but consist principally of one or more of the minerals plagioclase, pyroxene, and olivine.

Individual stones tend to be equidimensional in shape, though some may be angular owing to fragmentation by collision or terrestrial impact. Sometimes they are conical or dome-shaped (like the Apollo command modules) owing to a constant orientation when traveling through the atmosphere, so that one side suffers considerable heating with consequent loss of material. The fusion crust of stones is thicker than that of irons, often black, and may be dull or shiny. It may have a fluted or furrowed form caused by flow of molten material from the front to the back of the meteorite during flight through the atmosphere.

Tektites

2 ins

Stony iron meteorite

Iron meteorite

213

The interior of stones is usually gray or dark gray, granular in texture, and round chondrules may be apparent. Some nickel-iron metal may be evenly disseminated throughout or occur in occasional patches. They are difficult to recognize in the field because of the similarity in texture to terrestrial rocks.

Summary of features for recognizing meteorites

Meteorite falls are so infrequent there is no point in going especially to search for them, unless one has been reported to have fallen in a particular area. If you think that you have found a meteorite, however, the main features to look for in hand specimen are: firstly, the presence of chondrules; secondly, the presence of fusion crust; and thirdly, the presence of nickel-iron alloy, either comprising the whole meteorite, or as finely disseminated, shining grains or patches on a freshly broken surface. If you have recovered an object that was seen to fall from the sky, then it may be a meteorite, and it should be submitted to a museum or university for expert examination.

Tektites

Tektites are small glassy objects which, unlike meteorites, are found only in certain, rather limited areas of the Earth's surface. They are named according to the area in which they are found and the principal types are: *australites* from the southern part

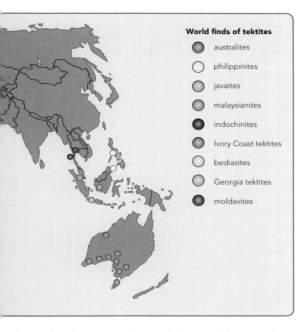

World finds of tektites

- australites
- philippinites
- javaites
- malaysianites
- indochinites
- Ivory Coast tektites
- bediasites
- Georgia tektites
- moldavites

of Australia, Tasmania, and coastal islands; *philippinites* from the Philippine Islands and southern China; *javaites* from Java; *malaysianites* from Malaysia; *indochinites* from Thailand and Indochina; *Ivory Coast tektites* from the Ivory Coast, West Africa; *bediasites* from Texas, USA; *Georgia tektites* from Georgia, USA; and *moldavites* from the Czech Republic.

It has been estimated that something like 650,000 tektites have been collected, of which the philippinites account for some 500,000.

Tektites are usually small, the majority being less than 11 oz in weight, and about half an inch to one inch across, but some examples up to 26 lbs are recorded. The shape of tektites is very variable, but diskoid, lensoid, button-shaped, tear-drop, dumb-bell, spherical and boat shapes commonly occur.

Some tektites are smooth and shiny but others have a rough, strongly etched and abraded surface, often with a system of grooves which reflect flow patterns within the glass. Most tektites are jet-black but thin flakes are transparent or translucent in shades of brown. The moldavites, however, are dark green, and in thin flakes are transparent and bottle-green.

Chemically, tektites comprise a silica-rich glass which is also rich in alumina, potash, and lime, and can be matched by a few igneous and sedimentary rocks. This has led to theories for the origin of tektites by the melting of terrestrial rocks through the impact of large meteorites or comets with the Earth. Other theories proposed an extra-terrestrial origin, but a terrestrial origin is now favoured.

Fossils

Fossils are the remains of animals or plants that are preserved in the rocks. It is very unusual for complete organisms to be preserved and fossils usually represent the harder parts of organisms, since these are the most resistant to decay and erosion. Most fossils therefore consist of the bones, shells, and tests of animals, and the leaves, seeds, and woody parts of plants. Moreover, even the original hard parts may not be preserved intact, because biological processes like scavenging, and geological processes like waves and storms, tend to disarticulate, break, and scatter all the different parts of the original organism. It is more common to find separate bones than a whole vertebrate skeleton, and the individual plates of crinoids are far more common than whole crinoids. Another factor is that animals or plants may shed parts of themselves in the normal course of their life: leaves fall from trees, seeds float or are blown away from the parent plant, and many arthropods, like trilobites, molt their carapaces as they grow. Also included as fossils are *trace fossils*, which are footprints, impressions, burrows, and borings into rock, left by organisms which are not themselves preserved. There are two examples in this book, the sponge borings of *Entobia* (*page 225*) and the bivalve borings of *Teredo* (*page 262*).

Fossils are found in most sedimentary rocks. They are particularly common in limestones, marls, clays, siltstones, mudstones, and shales, but they are less common in sandstones, conglomerates, and graywackes. Fossils can also be preserved in sedimentary ironstones and in volcanic ash. The majority of fossils are aquatic animals (i.e. they lived in seas, rivers, lakes or estuaries), because conditions for preservation are usually better in aquatic environments than on land. Even terrestrial animals and plants are more likely to be preserved in aquatic sediments, either through drowning (in a flood, for example) or because their bodies fell into water, or

were swept into it by floods and other sudden events. In this way, fossil land mammals are often found in the same deposits as fish, crocodile, and turtle remains.

How fossils are formed

The most common way in which fossilization begins is for the original shell or bone to be buried by sediment. Even the hardest parts of an organism will be broken down or dispersed after death if they are exposed to scavenging animals, bacterial action or the weather; so in general, the faster the rate of burial after death, the greater the chances of preservation. Organisms that burrow in sediment (like some bivalves and sea urchins) are effectively buried in sediment from the outset. Once buried, the surrounding sediment will usually take up an impression of the fossil. If this is left undisturbed as the rock becomes hard (i.e. lithifies), it will eventually become an external mold of the fossil. If the shell is a hollow structure, the internal cavity will often become filled with sediment or minerals, creating an internal mold of the fossil. In this way, a disarticulated bivalve shell will leave an external concave mold of its convex outer surface, together with an internal convex mold of its concave inner surface. (Take a modern beach shell, and try this out for yourself, experimenting with plasticene, modelling clay, plaster or latex.) If the shell has a completely enclosed interior cavity (as in brachiopods when their two shells have remained together after death), the internal mold will have the appearance of an intact brachiopod, but the surface of the internal mold will bear the impressions of the internal surfaces of the shells, not the outer surfaces.

Most organic material, other than tooth enamel, however hard it may seem, will change through geological time, so leading to two other important fossilization processes: replacement and formation of replicas. In the case of replacement, the original material is changed into other minerals, resulting in an intact fossil looking very similar to the original but made of a completely different material. This is particularly true of original aragonitic shells which are relatively unstable and soon change into calcite (*page 65*). Instead, the hard material is sometimes dissolved away completely, and the resulting cavity fills with sediment or minerals – or a combination – to produce a replica of the original fossil. Various combinations of all these processes also occur.

Sometimes the material in which a fossil is preserved lends it extra beauty, as in the case of pyrite, and opaline silica. Scientifically however, what is most valued is fine detail, preservation of softer tissues that are not usually preserved, and preservation of entirely soft-bodied organisms. Probably the best known and the most spectacular of these more complete modes of preservation, is that of insects and spiders, in amber (*fossil resin: see page 279*). Mummification of entire bodies can occur in very hot and very cold dry climates. Peat bogs and tar pits may preserve whole vertebrates complete with their skin and fur. Permafrost conditions have preserved occasional mammoths through refrigeration. If sediment burial is really sudden, softer fleshy tissues may be prevented from complete decomposition and decay, and many of their details will be preserved as films of carbonaceous matter. All these kinds of more unusual and complete modes of preservation have proved vital in giving us rare detailed insights into ancient life, and for this reason have been called Fossil Lagerstätten (fossil "ores", or, in effect, fossil treasures). Undoubtedly the best known example is that of the fossils, mostly of long-extinct arthropods, in the Middle Cambrian Burgess Shale in Canada. The entirely soft-bodied Ediacara organisms of the late Precambrian of South Australia, Canada and elsewhere are also very important, and include such unlikely fossils as jelly-fish and sea-pens along with others which are more enigmatic.

Names

Taxonomy is the science of naming and classifying organisms. Many living organisms bear a popular name as well as a formal scientific name. Although popular names may be well known and some are relatively precise, they are not really sufficient for consistent, accurate, scientific use. This is firstly because the same name may be used to refer to quite different animals or plants in different areas; and secondly because there are often different names for the same organism according to the language used. To overcome this problem, every known organism, whether modern or fossil, has been given a formal, internationally recognized, scientific name which is in Latin, though the Latin name is often an adaptation of words and names in another language.

To minimize ambiguous use of the formal name, one or a small group of specimens of every newly named organism is designated as a "type". Type specimens are deposited in a museum where other people can also study and compare them, and the author of a new name (who is often, but not always, the person who also discovered that organism) publishes the descriptions and illustrations of the type specimens in a proper book or scientific journal. This routine is intended to make it as clear as possible which particular organism bears a particular name. In effect, type specimens are the standards against which people can compare a name they have used for their own specimens or observations. Sometimes however, there are still problems about the use of a name, often for historical reasons, and considerable taxonomic research on all the relevant specimens and publications is needed to disentangle such difficulties.

The procedure for giving an organism its name is supported by internationally agreed rules of nomenclature, to ensure that everyone uses the same approach. Every different animal and plant has its own species or trivial name and that species has then also been assigned to a genus (pl. genera). Closely related species are placed in the same genus. The fullest, most precise, formal way of quoting an organism's name is to give its genus followed by its species name, followed by the name of the author who first described and named that particular organism, and finally, after a comma, the year in which that author published the name. The full name of our own human species is *Homo sapiens* Linnaeus, 1758. Note that the generic and species names (but not the author and date) should always be in italics (or underlined when writing and typing) because they are in Latin. Generic names begin with a capital letter, and can be used on their own (as in this book). Species names never begin with a capital (though you may see this in older publications when a species has been named after a person or place) and cannot be used on their own. They must always be preceded by the correct generic name or an unambiguous abbreviation (e.g. *H. sapiens*). This is because the same species name may have been used (validly) within many different genera, but taxonomic rules require that the combination of genus and species names must be unique to that one organism. Here, most of the illustrated fossils have been named at genus level only.

Higher level classification

As mentioned, closely related species are placed in the same genus, but also, closely related genera are in turn placed in a still higher level group (family), and so on, through further levels of classification. The sequence of main categories is: species, genus, family, order, class, phylum (pl. phyla) and kingdom, though these are often further subdivided and grouped to provide additional levels (e.g. superfamily), and extra categories are used for some groups of organisms in particular. In effect, this hierarchical system of naming and classifying organisms is a way of setting out how different organisms are thought to be related to one another, though many are scarcely more than convenient groupings of similar-looking organisms (broad similarity of appearance is not necessarily an indication of relatedness).

The formal names given to groups at these higher levels of classification are based on Latin and ancient Greek, but it is also common to anglicize them for informal use, as here (e.g. Molluska – mollusks). Research on names and relationships continues all the time, so classifications change too in the light of new studies.

No formal comprehensive higher-level classification has been used for the fossils shown here, as many groups of organisms have only a poor or non-existent fossil record, or their fossil representatives are relatively rare. However, the fossils are grouped here firstly according to kingdom (plants; animals), and within these, we have mostly used phyla (e.g. sponges; corals as representatives of the coelenterate phylum; mollusks). Below this level, we have grouped genera into one or more series of headings, partly based on formal classification (e.g. the headings: "trilobites"; "edrioasteroids"; "lycopsids"), and partly on convenient, non-taxonomic criteria like geological age (e.g. "Jurassic ammonites"; "Coal Measure Plants").

What is a species?

The most common definition of a species is a group of individuals that has the potential to interbreed freely to produce fertile offspring. Conversely, true species should be unable to breed successfully with members of other species. In a very few animal species and, more commonly, in some plant species, boundaries seem to break down, allowing successful breeding between members of different closely related species. These are exceptions, however, and successful breeding is not possible between members of the large majority of different species. In the case of fossils it is obviously not possible to test for interbreeding, and even in many living organisms, interbreeding potential is not known, or not easily tested. In practice, the most common approach is therefore to define a species on the detailed similarity of form and anatomy, and sometimes also lifestyle and behavior, shown by a group of individuals occurring in the same place at the same time (a natural population). For fossils, this place should be a single geological layer within a single small study area, with evidence that the fossils were once a living assemblage, rather than, say, an aggregation of shells brought together from many different places by currents and waves.

Collecting and studying fossils

Although many people acquire fossils by buying them, or being given them, it is also a very popular past-time for people to collect them in the field for themselves. Although you can do this on your own, it is usually better, at least initially, to make contact with other people with similar interests. This is a good way to learn more, but you will also find out about the increasing conservation and legal measures that can affect your interest, directly or indirectly. The main ways of making contact are natural history and geological societies, museums, libraries, and interpretative centers in conservation areas, and national parks. All of these either organize, or can tell you about, relevant activities in the immediate area. Some universities, colleges and schools also run extra-mural classes or informal groups, and provide facilities for non-specialists to learn more about geology and other field sciences. You will also find parallel organizations at a regional or national level. Access to the internet will provide another means of making contacts and obtaining further information, and there are also an increasing number of fossil identification sites. Using all these various approaches, you should be able to make contact with other people with similar interests, find out about field trips, and discover the best places to collect, as well as being able to learn more about fossils (and other aspects of geology) from talks and lecture courses given by enthusiasts and experts. Many organizations also arrange longer distance excursions at a reasonable cost.

This is also the best way to learn about the practical aspects of fossil collecting, ranging from recommended field clothing (according to the area, climate, and season), suitable items to protect yourself from injury, as well as the kinds of tools and other equipment you will need. Such organizations also emphasize adherence to standard published codes for the countryside and for considerate, safe and sensible collecting. Many fossiliferous places are now protected sites where collection is forbidden or restricted ("no hammering" – collect from scree and waste only), while others are industrial sites like working quarries which require special permission for entry and collection. If you intend to visit another country, you will often find that permission is also needed

beforehand to collect and take fossils out of that country. Your local group should be able to help you with all these aspects. Similarly, if you intend to travel some way from your own area, it is also useful (sometimes essential) to contact groups and organizations in that area too.

Although a few fossil finds may be worth considerable sums of money, and even common fossils can be traded at modest prices, many fossil collectors are motivated more by the scientific interest of their finds. Everyone likes to think that sooner or later they will make a special find of something rare or new, but it is worth remembering that even a carefully assembled collection of more common fossils may prove to be of considerable scientific value. One reason for this is that most collecting sites do not exist forever. Quarries are filled in or made otherwise inaccessible, outcrops can be washed away, overgrown, built upon, or drowned by reservoir schemes. Some popular and famous localities have simply become cleaned out. Other sites may be only temporary, because of a construction project.

A completely different factor is that in the course of scientific progress, specialists often need common fossils for data as well as rare ones, and it may also turn out that there are indeed new species in a collection of what seem like common fossils. Quality of preservation also varies in individual fossils, and amongst your collection you may have found a particularly well-preserved example of a common fossil which shows details not previously recorded. Even if none of these scientific aspects of your collection apply, a good collection may prove valuable as a learning resource for yourself and others.

For any of these reasons you may eventually wish to deposit some or all of your collection in a museum or other responsible institution with suitable storage, in which case it is essential that you have cared for it well. In fact, many museums will only take collections that are useful, and do not wish to be seen as convenient places to off-load unwanted material on the off-chance that someone will eventually take an interest in sorting it out later. Many fossils will disintegrate or become damaged if not packed carefully when collected, or stored properly afterwards. You should also keep a record of all relevant details of your specimens, a summary of which should be kept on a label with each individual specimen. At the very least, you need a field notebook to record your finds, simultaneously labelling them provisionally in the field as you make them, ensuring that your specimen labels and notebook match each other. Also do include any other relevant information and observations, and do keep your notebooks as they are really an archive. Your fossils should then be stored and documented (i.e. curated) in an orderly fashion, whether or not you also keep some on display. The widespread use of home computers makes documentation and labelling much easier now, but remember to back up your files properly as you go.

You may well wish to make further study of your fossils. This book provides only an entry point into this by helping you to make initial identifications. For more precise names, and for studying other aspects of your fossils, you will need to approach experts, to refer to more detailed books like monographs, and to make use of carefully identified museum collections, as well as laboratory facilities like microscopes and databases.

Societies, museums, and other centers should be able to advise you on all these aspects of collection and study, and in some cases, even provide facilities to help with this.

Why do people study fossils?
Although people collect fossils simply because they are attractive and interesting objects, fossils have also played a key part in the history of science, in our understanding of both the living world (biosphere) and the Earth as a planet – and they continue to do so. Fossils give us knowledge of life in the past, the study of which is called paleontology. Paleontology is unusual in being one of the few branches of science to which a non-specialist, non-professional can make a real contribution. Even so, this subject may seem a very quiet, esoteric backwater, apart from brief moments when headlines announce another spectacular dinosaur find. Appearances however are misleading. Every now and again,

paleontology has come to the fore as part of a fundamental debate at the very edges of our understanding of ourselves and our universe.

In the first place, fossils reveal that life on Earth has continually changed, and by implication, it will continue to do so. We may take this astonishing fact for granted now but this was not always so in the past. Awareness of our own extinct fossil relatives is due entirely to paleontology. More generally, the changing pattern of life through time provides strong supporting evidence for evolution, an idea that has won universal scientific acceptance in modern times, only after numerous long-running and heated debates that have at one time or another challenged governments, states, religious leaders, and distinguished scholars and scientists. Fossils (and other evidence) also reveal that environmental conditions on Earth also change continuously, an idea that was first realized when marine fossils were found high up in mountainous areas, showing that there must have been large-scale relative movements of sea and land through geological time. All these ideas helped to establish that the Earth was a dynamic, evolving planet and that it must also be very old.

Gradually, as geological evidence accumulated, it became clear that there was a relationship between the processes of rock formation and the fossil record, namely that successive layers of sedimentary rock contained their own unique groups of fossils. This is because, over geological time, new life forms emerge, and others go extinct, and wherever one looks at a series of fossiliferous rocks, this succession of life is always the same. Once a particular form has evolved it never repeats itself. There are of course numerous cases of broad similarities between different, relatively unrelated organisms, as in the helically-spired ammonite, *Turrilites* (*page 257*) and gastropod, *Microptychia* (*page 240*), but closer inspection reveals that the features of *Turrilites* are otherwise identical to that of other ammonites, and that this is therefore an example of convergence, not repeated evolution.

The succession of life provides the

foundations of biostratigraphy, the use of fossils to give a relative age for rocks. To gain an idea of how this works, you can discover from this guide that mammals are geologically younger life forms (Mesozoic to Recent) than trilobites and graptolites, which are much more ancient (Palaeozoic), and now extinct. So clearly, a rock containing trilobites must be relatively older than one containing mammal remains. Geologists developed this principle to work out a formalized scheme, the geological column (page 328), with its now familiar period-names like Cambrian and Jurassic, which gives relative ages based on characteristic fossils. The same principles are also used on various finer time-scales than this, to subdivide these main divisions into finer time units such as epochs, stages and, on the finest scale, sub-zones.

These fossil-based developments were critical, along with other geological evidence, in the growing realization that the Earth and hence also the Solar System, were much older than was previously realized. Some of this was based on guesses of how much time it must have taken for all the different organisms to have evolved, as well as for geological processes like sedimentation and mountain building to take place. By the mid-20th century, proof of the great age of the Earth (now put at around 4,500 million years) came from the use of chemical methods of obtaining a direct and numerical age of a rock. Most of these methods (radiometric dating) use naturally occurring radioactive isotopes found in the minerals of different rocks. The isotopes behave like a geological stopwatch which starts to run when the crystals of a mineral first form. As time proceeds, more of the original unstable isotope is transformed into another related, but stable, isotope. The proportion of the amounts of the unstable to stable isotopes in a mineral therefore slowly decrease through time, and, since the rate of this change (radioactive decay) is regular and known, the proportion can be used to work out how much time has passed since the mineral first formed. This gives an absolute age (ie. a known figure in thousands or millions of years) for the particular rock containing that mineral sample, though the geologist has to

check from other evidence that the mineral really did crystallize at the same time that the rest of the rock formed.

Vital though these and other physical-chemical methods have proved for giving precise and absolute ages for the geological column, and also for dating rocks without any fossils (as in most of the Precambrian), fossils are still the least expensive and most readily applied method of dating rocks from the Cambrian onwards. Fossils are still widely used in the search for Earth resources like water and fossil fuels, and in almost any other instance where a geological map or section is needed. The whole approach requires that fossils should be identified accurately and this is a key part of many paleontologists' work in museums, geological surveys, universities, and in industry. Many macrofossils (notably trilobites, graptolites, and ammonites) can be used to give an age quite quickly, sometimes even in the field by an experienced specialist, but microfossils such as plant spores, foraminifera and conodonts are preferred. Although these require specialist techniques (which is why they are not included here), this is offset by the fact that large numbers of samples can often be collected easily. In general, age-diagnostic fossils should be widespread in their distribution and should have features which evolved relatively rapidly, so giving the highest resolution. (The best resolution from fossils alone is about 250,000 years, other than for the Quaternary, where resolution is sometimes finer.)

Although nearly everyone is now aware of the possibility of global climate change in our own time, perhaps due to human activities, paleontologists and geologists have long known that global change is also a natural fact of Earth history on time scales ranging from thousands of years to millions of years. Much of the future has already been written in the past, in rocks and fossils. Some of the evidence for environmental change is based on analogy of fossils with their living counterparts (as a very obvious example, the occurrence of a fossil fish must signify an aquatic environment). Other evidence comes from comparison of details of morphology between different kinds of organisms and their function, as when we conclude that the flippers of extinct marine reptiles, like those of modern seals, whales, and turtles, must indicate a swimming habit. Many fossil skeletons also have chemical signatures indicating the temperature and salinity of the water in which they lived, or aspects of their diet. The geographical distributions of fossils (paleobiogeography) provided early evidence for the spectacular idea of continental drift, now emphatically proven using physical methods. Fossils help to reconstruct past climates, the nature and relative depth of the sea floor at a particular time and place, and the patterns of currents and other features of ancient oceans. At the same time the paleontologist finds out how life has evolved, ecologically, and how it has responded to all these enormous changes. Paleontology has been at the center of important debates such as why fossils became common suddenly in the earliest Cambrian, the ways in which animals and plants first colonized the land and why dinosaurs became extinct.

More recently, the accumulation of data from fossil collections has allowed a great deal of quantitative analysis to be carried out on the global fossil record. The best-known result of such work is the revelation of "mass extinctions", those mysterious and threatening events in the history of the Earth when a substantial proportion of all the life forms then existing were wiped out, probably through drastic and relatively sudden changes to the global environment, or perhaps due to a massive meteorite impact. The role of the carbon dioxide cycle and the greenhouse effect are an important part of this. Palaeontology gives a key to these processes because organisms are responsible for limestone, oil and coal, and large amounts of carbon are locked up in the Earth's crust when these deposits form in abundance. Carbon is released when these are eroded and when we ourselves burn fossil fuels. This leads naturally to the idea that Earth and life are intimately linked together environmentally, and evolutionarily (cf. Gaia), since carbon dioxide is also generated by volcanism, while the oxygen on which all animals depend is almost entirely the product of plant-life. It is a measure of how

paleontology has entered people's awareness that many of the major issues mentioned here have at one time or another been the subject of books, novels, plays, and films.

Scope of this guide

There is not sufficient space in this guide to treat all known fossil species as this would run into millions, and it is not even possible to cover every genus or family. The fossils shown here are included because they are mostly common or widespread, or because they are otherwise typical of higher level groups which are common or widespread. We have concentrated on macrofossils (fossils that can be easily seen in hand specimen) and omitted microfossils, the collection and study of which usually requires specialist techniques. Choice of examples also reflects the fact that animals are much more abundant as macrofossils than plants. Most people who have a general knowledge of geology and paleontology are familiar with the spectacular, large, extinct vertebrate fossils such as dinosaurs, marine reptiles, and some of the larger mammals like mammoths and sabre-tooth cats, as well as our own human fossil relatives. Children in particular like to learn about the names and habits of these fossils. Once you start looking for fossils however, you soon realize that the chances of finding such fossils are very small, and in most cases, whole skeleton finds are even rarer. For this reason the fossils in this guide are mostly invertebrates, and the relatively few vertebrate examples that we have included are typical of the incomplete finds that people are most likely to make (e.g. teeth). If you wish to learn more about these extinct animals, there are plenty of well-illustrated popular books about them. Amongst invertebrate macrofossils, mollusks are the most important and have been given the most space (*pages 236 to 269*). At the other extreme, insects are relatively rare, notwithstanding their huge richness and abundance as living organisms, and only a few examples are given.

It is important to be aware that some of the fossils shown here are really single examples of a set of closely similar and related forms, each of which has its own generic name, though they are difficult to distinguish from each other without specialist study. Research continues all the time, names and classifications change, and it sometimes becomes difficult to say what the correct generic name is for the fossil shown. In these cases, we have used generic names rather loosely to cover a closely related group, more in the manner of family or subfamily groupings, than a single strict genus. For some of these "genera", we have also used inverted commas, e.g. "*Pleurotomaria*", and in many places we have also mentioned close relatives in the descriptions.

The age ranges given here can be compared with the geological column on page 328. In this book, "Recent" is used for living organisms together with their immediate fossil record going back to the last 10,000 years. The geographical range is given with abbreviations as follows: NA – North America; SA – South America; E – Europe; Af – Africa; Aust – Australasia. Asia is not abbreviated. "Worldwide" indicates probable occurrence in all these regions. Where a geological age range is uncertain, a question mark is used next to the age range in the text.

Contributors to the fossils section

The original text, written by Dr W. R. Hamilton, has been revised and updated by specialist contributors, listed below, under the direction of Dr Brian Rosen. All contributors are from the Department of Palaeontology, The Natural History Museum, London (unless otherwise shown). In addition, all sections have been further amended by Dr Brian Rosen.

New photography

Frank Greenaway (Photographic Unit of the Department of Exhibitions and Education, The Natural History Museum) provided new photographs for the following:

Actinocyathus	Anonaspermum
Aturia	Cheiracanthus
Chlamys	Coccosteus
Endoceras	Eutrephoceras
Favosites	Fissidentalium
Hamites	Libellulium
Modiolus	Neohibolites
Orthoceras	Osteolepis
Productus	Carnallite (see page 60)

Introductory text

New text by Brian Rosen, with assistance from Paul Taylor

Arthropods – Trilobites

Revised by Richard Fortey

Arthropods – Insects

Revised by Andrew Ross

Bivalves

Revised by Noel Morris, Phil' Palmer, John Cooper, Jonathan Todd, Paul Jeffery, and Andrew Gale (University of Greenwich)

Brachiopods

Revised by Sarah Long, with assistance from Robin Cocks, Ellis Owen, and Howard Brunton

Bryozoans

Revised by Paul Taylor

Cephalopods

Revised by Michael Howarth, Neale Monks, and Phil' Palmer, with assistance on belemnites from Peter Doyle (University of Greenwich). Neale Monks provided a new section on nautiluses.

Corals

Revised by Brian Rosen, with assistance from Jill Darrell, and Colin Scrutton (University of Durham)

Echinoderms

Revised by Andrew Smith and David Lewis

Fishes

Revised by Sally Young and Alison Longbottom

Fossil land plants

Revised by Paul Kenrick and Tiffany Foster

Gastropods

Revised by Noel Morris, Phil' Palmer, Jonathan Todd and John Cooper

Graptolites

Revised by Claire Mellish and Richard Fortey

Mammals

Revised by Andy Currant, Peter Andrews, and Jerry Hooker

Reptiles and birds

Revised by Angela Milner

Scaphopods

Revised by Phil' Palmer

Sponges

Revised by Steve Donovan, Paul Taylor, and John Cooper

Sponges

The simplest, multicellular animals. Structure usually radial with a central cloaca and surfaces covered with pores. Several forms are shown here that have characteristic shapes but the detailed identification of many sponges relies on the study of thin sections.

Chenendopora *Cretaceous: E*
Medium-sized to large, usually 2 to 4 inches high. A vase-shaped sponge with a large, wide cloaca. Pores on outer and inner faces more clearly visible inside cloaca. Attachment stem shown at base.

Siphonia *Middle Cretaceous–Tertiary: E*
Globular, widening downward. Cloaca narrow, less than 0.5 of an inch. Surface generally smooth with small pores. Stalk long and slender.

Ventriculites *Cretaceous: E*
Thin walled, vase-shaped, high to flattened and saucer-shaped (both shown here). With strong vertical grooves on the outer surface marking the course of canals and large pores on the upper face. Cloaca varying in width with shape of whole animal.

2 ins

Chenendopora

Entobia

Peronidella

Siphonia

Ventriculites

Ventriculites

224

2 ins

Hydnoceras

Doryderma

Peronidella *Triassic–Cretaceous: E*
A medium-sized form consisting of numerous cylindrical units each less than 0.5 of an inch in diameter and radiating from a common base. Each has a small cloaca at its tip.

Doryderma *Carboniferous–Cretaceous: E*
A relatively large, branching, plant-like sponge. Branches at least 0.5 of an inch in diameter.

Hydnoceras *Devonian–Carboniferous: NA E*
Small to large (shown here). Vase-shaped. Surface with a network pattern formed by large vertical and transverse ridges with finer ridges between them. Regularly arranged swellings are present, usually at the intersection of large ridges; these swellings delimit the eight faces of the sponge. This genus represents a group that is particularly common in the Devonian of New York State.

Entobia *["Cliona"] Jurassic–Recent: Worldwide*
A small boring sponge that forms nodular swellings in shells or on rock surfaces. These swellings are joined by slender connecting rods. These are trace fossils usually preserved as casts (here) or molds, which, though produced by clionid sponges, are known by the trace-fossil name *Entobia*. Here, the infill of the boring is preserved, but the bored shell has been dissolved away.

225

Corals

Skeletons calcite or aragonite and commonly preserved as fossils. Detailed classification is based largely on details best seen when specimens have been cut or thin sectioned and examined by light and scanning microscopes. However, natural weathering and breakage of corals reveal useful details, and identifications can also be narrowed down by taking into account the age of the deposits in which the corals are found.

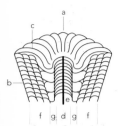

Generalized morphology of single corallite

Corals in transverse and calicinal views consist either of a single star-like radial structure *(corallite)*, as in *solitary* corals (like *Montlivaltia* here) or of repeated patterns of radial structures, as in *colonial* corals (like *Hexagonaria* here). The outer, youngest, or *oral* end of a corallite is the *calice* (a), occupied in life by an anemone-like polyp. Corallites usually surrounded or supported by a wall (b) *(theca)*. If walls are missing in colonial corals, corallites consist only of *"centers"*. Corallites may be joined by intercorallite tissue *(coenosteum)* (like *Favia* here). Radial elements (c) *(septum, pl. septa)* consist of plates or spines sometimes extending beyond corallite walls as *costae* (like *Montlivaltia* here). *Axial structure* (d) and/or rod- or plate-like columella (e), present or absent. Longitudinal sections reveal internal supporting structures like small blistery plates (f) *(dissepiments)*, and broader plates (g) *(tabula, pl. tabulae)*. Colonial forms commonly *massive* (head coral), or branching, encrusting, platy, tabular or columnar. Branching forms may consist of a single corallite for each branch *(phaceloid* branching) as in *Siphonodendron* or branches composed entirely of numerous corallites *(ramose* branching) as in *Acropora*.

Favosites

2 ins

Tabulata

Ordovician–Permian
Always colonial. Corallite diameters usually small, typically less
than 0.04 to 0.2 of an inch. Walls often with pores. Septa absent,
or very few, rarely more than 12; often as longitudinal series of
spines or spinose combs. Tabulae usually well developed.

Favosites *Upper Ordovician–Middle Devonian: Worldwide*
Colonies massive, tabular to domal and subspherical. Corallites
prismatic, in close contact. Walls thin, with mural pores. Septa as
longitudinal series of spines, or absent. Tabulae complete, subhorizontal.

Coenites *Middle Silurian–Middle Devonian: NA E Asia*
Colonies, small, delicate, ramose branching. Corallites very small,
prismatic, in close contact, opening at surface at acute angles,
with crescentic calices. Walls with sparse pores. Septa or septal
combs in one to three longitudinal rows. Tabulae thin, transverse
to inclined.

Halysites *Late Ordovician–Upper Silurian: Worldwide*
Colonies in transverse view have corallites arranged in chain-like
ranks (a) (uniserial rows), connected to each other in networks.
Ranks palisade-like, with spaces (b) *(lacunae)* between them.
Corallites rounded to elliptical, with quadrangular tubules (c)
alternating between them. Walls thick. Septa weak to absent.
Tabulae (d) horizontal, in both corallites and tubules.

Halysites: longitudinal section

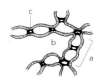
Halysites: transverse section

Halysites

Favosites

Coenites

Halysites

2 ins

227

Aulopora *Ordovician–Permian: Worldwide*
Colonies of net-like series of attached diverging corallites connected singly to each other, and usually adhering to a surface (*reptant* habit). Corallites conical or barrel-shaped, inclined away from surface. Walls moderately thick. Septal spinules present or absent. Tabulae usually absent or sparse, oblique.

Syringopora
Upper Ordovician–Lower Carboniferous: Worldwide
?Upper Carboniferous: Worldwide ?Lower Permian: NA
Colonies phaceloid, with irregular subparallel branching cylindrical corallites connected by tubules. Walls moderately thick. Septa spinose to absent. Tabulae numerous, sagging, funnel-like.

Rugosa

Middle Ordovician–Upper Permian
Solitary or colonial. Corallites may attain diameters of more than 1 inch, even up to 4 inches or more in solitaries. Corallite symmetry fundamentally bilateral, though often radial in appearance. Septa usually numerous, typically in two alternating radial lengths (*major* and *minor* septa). Internal structures often include well-developed zonal arrangement of dissepiments and tabulae, with dissepiments usually confined to marginal region, or absent. Axial structures sometimes well developed, with or without a rod or plate.

Syringopora

2 ins

Aulopora

2 ins

Hexagonaria

Palaeosmilia

Siphonophyllia

Hexagonaria *Middle–Upper Devonian: NA E Asia Aust*
Colonies massive to tabular, with closely packed corallites about 0.5 of an inch in diameter. Septa long, notably thicker midway along their length, bearing small ledge-like ridges (*carinae*) oblique to corallites. No axial structure. Dissepiments numerous in marginal zone. Tabulae flat to subhorizontal. Calices commonly with central depression, surrounded by outer platform corresponding to dissepiment zone, edge of which is often marked by an apparent "inner wall" in transverse sections. [Genus easily and commonly confused with *Prismatophyllum*, Lower–Middle Devonian: NA.]

Hexagonaria: transverse section

Siphonophyllia *Lower Carboniferous: E Af Asia*
Solitary, large diameter (greater than 2 inches), sometimes very long, with numerous bends. Septa numerous, withdrawn from axial region, which appears void-like. Dissepiments numerous, in well developed outer zone, around wide inner zone of numerous complete tabulae which are flat to slightly domed upward in axial region and turned down in marginal region. Septa thickened here in this down-turned zone. In mature specimens, marginal zone consists only of blistery dissepiments and septa are withdrawn from wall (*lonsdaleoid* condition).

Siphonophyllia: transverse section

Palaeosmilia *?Upper Devonian, Carboniferous: E Af Asia Aust*
Solitary; large (greater than 2 inches). Septa very numerous; majors extending close to axis. Dissepiments numerous, in well developed outer zone, around wide inner zone of numerous incomplete tabulae domed upward in axial region. [Colonial forms with similar structures should be referred to *Palastraea* Carboniferous: NA, E Af Asia.]

Palaeosmilia: transverse section

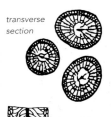

transverse
section

longitudinal
section

Siphonodendron

Siphonodendron *Carboniferous: NA E Af Asia Aust*

Colonies phaceloid, with subparallel branching cylindrical cora-llites, 0.08 to 0.5 of an inch in diameter. Septa strongly alternat-ing. Columellar plate lenticular in section, sometimes absent, sometimes continuous or nearly so with septa aligned in same plane. Dissepiments rarely absent, usually in one or a few margin-al rows, within which are well developed flat tabulae, pointing upward where they meet columella. [*Lithostrotion* Carboniferous: NA E Af Asia, Aust, is a closely related form with very similar internal corallite detail, but with close-packed (not phaceloid) corallites.]

Actinocyathus *Lower Carboniferous: E Asia*

Colonies massive, tabular, with closely-packed corallites about 0.5 of an inch in diameter. Septa withdrawn slightly from axial region and very strongly from walls, where marginal zone consists only of large blistery dissepiments (*lonsdaleoid* condition). Axial structure a complex of septal elements and small axial tabulae giving characteristic cobweb pattern. Tabulae between axial structure and outer zone are flat to sagging. [*Lonsdaleia* Car-boniferous: ?NA E Asia, is a closely related form with very similar internal corallite detail, but with phaceloid (not close-packed) branching corallites.]

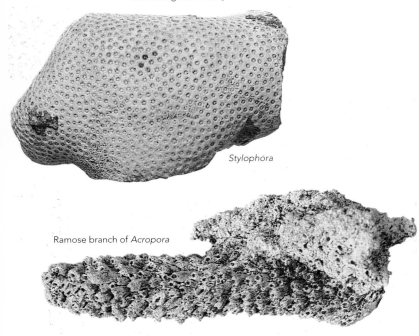

Stylophora

Ramose branch of *Acropora*

2 ins

2 ins

Siphonodendron

Actinocyathus

Scleractinia

Middle Triassic–Recent
Solitary or colonial. Colonies often elaborate in form. Corallites may attain diameters of 1 inch, even up to 8 inches or more in solitaries. Corallite symmetry fundamentally bilateral, though often radial in appearance. Septa usually numerous, arranged in successively smaller size orders, usually in multiples of 6 (e.g. 6+6 +12+24+48 etc). *Acropora, Porites, Favia,* and *Diploria* (all shown here) are important reef builders, especially since the Miocene.

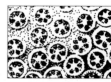

Stylophora

Stylophora *Paleocene–Recent: Worldwide in lower latitudes*
Colonies encrusting, nodular and branching ramose, up to 20 inches or more across. Branches narrowly cylindrical to stoutly lobate. Corallites small, about 0.04 of an inch in diameter, joined by narrow solid spinulose coenosteum. Walls often project strongly (except in worn fossils), often assymmetrically as "hoods", giving a rasp-like appearance to surfaces of unworn colonies. Septa 6, usually extending to axis where a slim rod may develop.

Acropora *Paleocene–Recent: Worldwide in lower latitudes*
Mostly ramose branching (e.g. bushy, stagshorn, and elkhorn forms), columnar or encrusting; table forms also common, developed by dense horizontal anastomosing of radiating branches, often with close short upturned equal branch-tips. (Overall colony form of fossil specimens is rarely seen however, as *Acropora* mostly occurs as broken fragments.) Branches usually cylindrical, tapering, often long, with strong *axial corallite* at each tip, from which surrounding *radial corallites* have been generated. Corallites usually project strongly as tubules or as pustules about 0.08 of an inch across, often turned obliquely toward branch tips. Coenosteum porous, with regular pattern of very fine flaky or spinulose elements. Septa few (12, 6, or virtually absent), spinelike, deep.

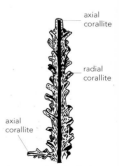

axial corallite

radial corallite

axial corallite

Acropora: longitudinal section

Septal scheme in Porites

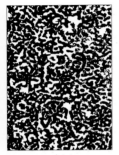

Transverse views

Thamnasteria: transverse views

Porites *Eocene–Recent: Worldwide in lower latitudes*
Colonies massive, encrusting, platy, columnar, nodular, or ramose branching, up to several feet across. Corallites small (0.04 to 0.08 of an inch in diameter), close-packed or joined by coenosteum, relatively inconspicuous, giving surfaces a finely pitted, cellular or grainy appearance. Skeletal elements discontinuous, porous, three-dimensional reticular structure of very small granulated pillars and rods. Septa bilaterally arranged, always 12, with characteristic plan, often with small pillars projecting along their length, especially around axial area. (Skeletal details, including corallites, difficult to differentiate in fossil forms, sections through which may appear as discontinuous, subparallel, undulating elements, or vaguely-defined centers; widely mistaken for sponges.)

Thamnasteria *?Middle – Late Jurassic: Worldwide*
Colonies platy, columnar or branching ramose. Corallites small (0.04 to 0.08 of an inch in diameter), densely packed. Septa of adjacent corallites continuous with one another *(confluent)*; longer septa converge at corallite centers; shorter septa curve to join longer septa. Septa bear small ledge-like outgrowths (pennulae). Longitudinal bars connect adjacent septa especially at margins of corallites where they make a wall-like structure. Columella a conspicuous small rod. (Confluent septal patterns occur in numerous other corals, some of which have been confused with *Thamnasteria*.)

Cunnolites *Late Cretaceous: NA E Asia Af*
Solitary, round to elliptical, discoidal to domed, usually 1 to 4 inches in longer diameter, with deep central groove and tightly-packed septa in very numerous, poorly-differentiated size orders. Septa perforate, bearing numerous flanged structures *(pennulae)* and connected by rods. Undersurface bears fine surface "skin" *(epitheca)* with closely concentric ridges. (This coral has also been widely known as *Cyclolites*.) Characteristic shape is similar to the unrelated modern mushroom corals like *Fungia* Miocene–Recent: Indo–Pacific.

Porites

2 ins

Isastrea *Jurassic: Worldwide*

Colonies massive, encrusting or platy. Corallites closely packed, polygonal to round, 1 to 6 inches in diameter. Septa in weakly differentiated size orders, often granular on upper edges, and with small flange-like ridges *(carinae)* oblique to corallites. Walls weak, discontinuous or absent, often with septa of adjacent corallites continuous with one another in wall zone *(confluent)*. Septa joined by transverse bars *(synapticulae)* in wall and axial areas. Dissepiments numerous. Columellar spongy, weak.

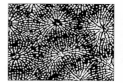

Isastrea: transverse view

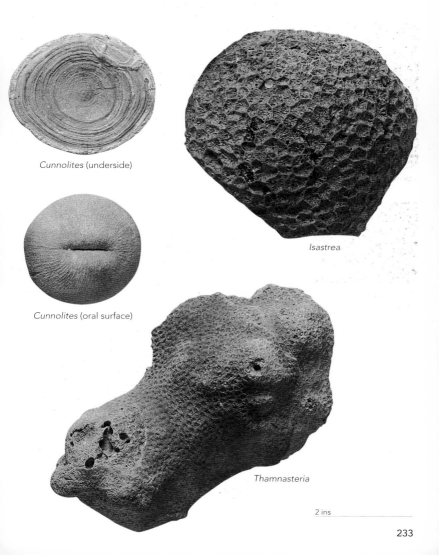

Cunnolites (underside)

Cunnolites (oral surface)

Isastrea

Thamnasteria

2 ins

233

Montlivaltia *Jurassic–Cretaceous–?Eocene: Worldwide*
Solitary, cylindrical to horn-shaped. Calices with elongate axial pit, without axial structure. Septa long, in numerous size orders; may appear almost smooth in sections of larger septa, but otherwise have regular, oppositely-paired, cuspate outgrowths along their length. Septa reaching axial area often slightly thickened at their ends. Outer parts of septa arch gently upward from axis, and descend as costae on external walls of corallite where, lower down, they are usually covered by strips of a fine surface "skin" of closely parallel ridges and grooves *(epitheca)*. Dissepiments numerous, strongly arched around margins.

Placosmilia *Cretaceous–Eocene: E*
Like *Montlivaltia* but having a flattened, elongated to meandering cross-section.

Thecosmilia *Jurassic–Cretaceous: Worldwide*
Like *Montlivaltia* but in phaceloid branching colonies. Corallites robust, large (greater than 1 inch in diameter), dividing internally to form new branches.

2 ins

Placosmilia

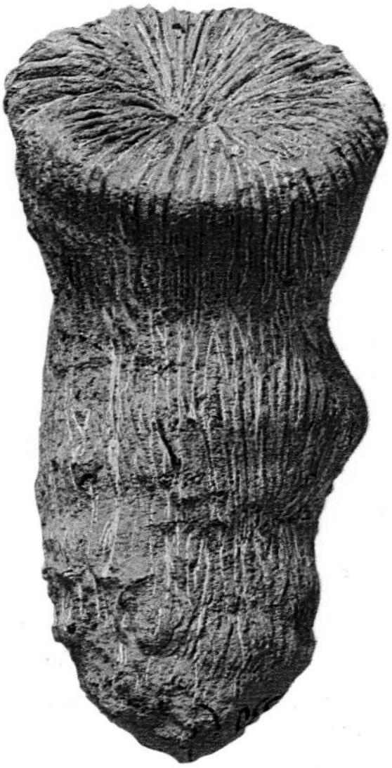

Montlivaltia

Thecosmilia

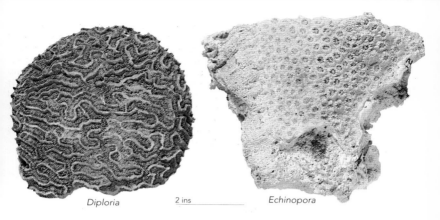

Diploria   2 ins   Echinopora

Favia *Cretaceous–Recent: Worldwide*

Colonies massive to encrusting or sometimes columnar, up to 3 feet or more across. Corallites 0.2 to 1 inch in diameter, typically about half an inch. Walls well-developed, thin, upward-projecting. Coenosteum narrow, of costae, thickened wall and dissepiments. Septa with serrated edges and granular-spinose surfaces; projecting upward over walls where sometimes thickened, then continuing as costae. Septal margins within calices steeply descending, sometimes making a crown-like structure around deep axial region. Abundant internal dissepiments. Axial structure weak to strong and spongy. New corallites arise by internal division of mature corallites.

2 ins

Favia

Diploria *Upper Cretaceous–Recent: NA E Caribbean.*

Characters of *Favia*, but colonies *meandroid* ("brain coral"). Walls very elongate, and meandering, enclosing fork-ended valley systems of numerous barely distinguishable corallite centers along their length. Coenosteum narrow to wide, sometimes as wide as valleys. Axial structure continuous along valley axes.

Echinopora *Miocene–Recent: Indo–Pacific*

Colonies nodular, encrusting, platy, irregularly columnar or in large delicate curving leaf-like fronds. Corallites round, projecting, 0.08 to 0.2 of an inch in diameter, with strong walls, joined by strongly spinose, tabular coenosteum. Septa in 3–4 well-differentiated size orders, and strongly spinose over well-developed wall. Axial structure large and spongy. Dissepiments numerous. New corallites arise from coenosteum between mature corallites. (This coral is similar to its well-known relative, *Montastraea*. Cretaceous – Recent: Worldwide.)

2 ins

Parasmilia

Parasmilia *Cretaceous–Recent: Worldwide*

Solitary, 0.5 of an inch in diameter; narrow inverted conical form, often curved with preserved point of attachment. Septa projecting slightly upward and continuing externally as fine costae. Calice deep. Axial structure deep, conspicuous, spongy.

235

Mollusks

The most numerically important group of fossil animals which includes three major groups, still living today: Gastropoda (snails), Cephalopoda (squids, nautiloids, ammonites etc.), and Bivalvia (bivalves or clams). Other groups included here are bellerophonts, now extinct, and scaphopods (tusk shells). Not included are the extinct bivalve-like rostroconchs, and the Polyplacophora (chitons or coat-of-mail shells). There are also non-shelled groups with no significant fossil record.

Bellerophonts

Gastropod-like shells, but it is not certain that the body plan is twisted like that of a gastropod. Terms used are as for gastropods (below).

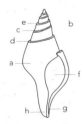

Bellerophon: aperture

Bellerophon *Silurian–Triassic: Worldwide*
Usually 1 to 3 inches wide. Shell wide, flaring near aperture (a). Bilaterally symmetrical; last whorl covers earlier whorls which are only visible in deep holes on either side. Front margin of aperture carries deep slit (s). Strong ridge (r) around middle of whorl, and growth lines strong.

Gastropods

The gastropod shell may be coiled (snails), uncoiled (limpets) or reduced (slugs). Important features of gastropods are related to the coiling, aperture, columella and shell sculpture. A *whorl* is a complete coil of the shell; the *last whorl* (a) is the largest and the *spire* (b) is all of the shell except the last whorl. The *suture* (c) is the line along which the whorls are joined. If the whorls are angular, then the main angle, where the shell turns inward toward the suture, is known as the *shoulder* (d), and the part above the shoulder is known as the *ramp* (e). The *aperture* (f) is the opening to the outside. Its shape and features of the lips are important. Sometimes the aperture is rounded but in other cases it may be produced below and folded over, forming an *anterior canal* (g). Less usually a *posterior canal* may be developed. The *columella* (h) is the central column of the shell *(clearly shown in Clavilithes on page 244)*. It sometimes bears ridges known as *columellar plications*. The columella may have a hollow center known as the *umbilicus*. A pad of *callus* is often developed in the columellar area. The sculpture of a gastropod shell may follow the line of coiling (i.e. *spiral*), or it may be parallel to the growth lines (i.e. *axial*). In the descriptions, "width" refers to the diameter of the largest whorl, and "height" to spire height.

Typical structure of a gastropod as shown by Clavilithes

Poleumita *Silurian: NA E*
Usually 2 to 4 inches wide. Upper surface flattened. Ornament of fine lamellae and slightly raised spines on shoulder. A similar form *Straparollus* (Silurian–Permian: Worldwide) has an aperture of different shape.

Poleumita: aperture

Trepospira *Devonian–Permian: NA SA E Af*
Usually about 1 to 2 inches long. Conical. Deep slit (s) on front edge of aperture. Faces of whorls flat; outer edge of whorl sharp. Aperture as shown. Surface smooth with row of tubercles just below suture; these distinguish *Trepospira* from *Liospira* (Ordovician–Silurian: NA E Asia) which has a completely smooth surface.

Trepospira: aperture

Mourlonia *Ordovician–Permian: NA E Asia Aust*
Usually 1 to 3 inches long. Conical; sutures more deeply impressed than in *Trepospira*. Two to three ridges along shoulder and just above suture on earlier whorls. Strong slit on front edge of aperture (not shown here).

Straparollus: aperture

Bellerophon

Mourlonia

Trepospira

Poleumita

2 ins

237

Worthenia: aperture

Worthenia *Carboniferous–Triassic: Worldwide*

Medium-sized, usually about 1 to 2 inches high. Shell relatively higher than in *Mourlonia*. Whorls angular with flattened faces and strongly ridged shoulder bearing small tubercles. Under surface of last whorl with spiral ridges crossed by strong growth lines, thus forming a network pattern. Aperture almost square with thickened back edge and small slit (s) on front margin. Umbilicus absent.

"Pleurotomaria" *Jurassic–Cretaceous: Worldwide*

Usually up to 4 inches long and/or 3 inches high. Coiling low (as shown here) to high as in *Bathrotomaria*. Umbilicus present. Aperture rounded with long slit on upper front edge (shown here just below green spot). Heavy ornament of large swellings on shoulder of whorls and near suture. A spiral band of different sculpturing lies between the two rows of tubercles. Spiral grooves and strong growth lines also present. (*Pleurotomaria sensu stricto* is known only from the Jurassic. There are other closely similar, related forms which have different ranges.)

Bathrotomaria *Jurassic–Cretaceous: Worldwide*

Medium-sized to large, up to 3 inches high. Closely related to *Pleurotomaria* and similar to flattened or high forms of that genus. Deep slit on front margin of aperture (shown here) which becomes filled in during growth as a strong spiral ridge visible almost as far as the apex. Ornament also includes numerous spiral ridges and grooves and weaker growth lines.

2 ins

Bathrotomaria

"Pleurotomaria"

Worthenia

Calliostoma

Cirrus

Platyceras   2 ins

Platyceras *Silurian–Permian: Worldwide*
Representative of a group (Platyceratoidea) in which the last whorl is very large and the other whorls are much smaller. The specimen shown is extreme and other members of the group may be more similar in general form to *Mourlonia*. Border of aperture may be wavy or straight. Ornament of growth lines. No slit on front margin of aperture.

Calliostoma *Cretaceous–Recent: Worldwide*
Medium-sized, usually 0.5 to 2 inches long. Conical with pointed, straight-sided spire. Aperture as shown. Umbilicus absent. Inner shell layer commonly like mother-of-pearl (shown here). Sutures may or may not be deeply indented. Ornament of spiral ridges, varying in distribution from area near sutures only, to covering whole surface of whorls, also varying in strength. No slit on margin of aperture.

Calliostoma: aperture

Cirrus *Triassic–Jurassic: SA E*
Medium-sized to large, 1 to 3 inches wide or high. Flattened to high conical (shown here). Umbilicus large, varying with height of shell. No slit on margin of aperture. Ornamentation of strong vertical ridges and weaker spiral ridges. Sutures shallowly depressed. Aperture almost circular. Coiling opposite in direction to most gastropods.

239

Ooliticia *Jurassic–Cretaceous: Worldwide*
Small to medium-sized, usually 0.25 to 2 inches high. Steeply conical with faces of whorls rounded. Umbilicus absent. Aperture diamond-shaped to rounded. Ornament of strong spiral ridges carrying tubercles and crossed by fine vertical ridges. No slit on margin of aperture.

Loxonema *Silurian: NA E*
Medium-sized up to 3 inches high. High, pointed spiral with whorls having rounded walls. Sutures deep. Umbilicus absent. No slit on margin of aperture but outer margin with a deep curved depression known as a sinus (s). Ornamentation absent.

Microptychia *Carboniferous: NA E*
Medium-sized, up to 3 inches long. High pointed conical. Sutures deep with ornament of short vertical ridges; these increase in strength upward and may completely cover the top whorls. Lower whorls smooth. Aperture almost circular and lacking sinus. Walls of whorls rounded but more convex near lower suture.

Loxonema: aperture

Natica *Paleocene–Recent: Worldwide*
Medium-sized, usually 0.5 to 2 inches high. Shape ranges from almost spherical (shown here) to conical. Walls of whorls rounded and sutures usually deep. Surface smooth and may be shiny with a few lamellar growth lines near aperture. Umbilicus usually present but columellar callus may cover it. Last whorl very large. Aperture oval to circular. Inner lip thickened, outer lip thin.

2 ins

Ooliticia

Natica

Loxonema

Microptychia

Calyptraea

Xenophora

2 ins

Xenophora Cretaceous–Recent: Worldwide

Medium-sized, up to 3 inches wide. Conical with flattened base. Last whorl with sharp outer margin. Wide umbilicus and characteristically shaped aperture (shown here). Inner margin thickened. Surface rough with depressions where shell fragments, and other foreign particles such as pebbles, were attached during life. Some shell fragments are still present on the right side of the specimen, and depressions on the upper side show the patterns of attached particles. Some species of *Xenophora* have lightly sculptured surfaces.

Xenophora: cross-section showing flattened base

Calyptraea Cretaceous–Recent: NA SA E

Medium-sized, up to 3 inches wide. Flattened to high conical shell consisting of a few wide whorls. Last whorl very large and lower surface deeply concave with a small internal shelf which has a twisted border (columella) and a small umbilicus at its highest point. Ornamentation of weak growth lines and occasional tubercles which are stronger near the lower edge.

Calyptraea: cross-section showing concave lower surface

241

Crepidula: cross-section

Crepidula (slipper limpet) *Tertiary–Recent: NA E*

Medium-sized, usually 1 to 3 inches long. Flattened, convex, and slipper-like. Whole shells consist of a single whorl. Under surface characteristic, having deep concavity as shown, and a wide concave shelf (s) which lacks the thickened columellar edge of *Calyptraea*. Ornamentation of ridges, spines, and growth lines may be present on upper surface.

Stellaxis (sundial shell) *Eocene–Recent: NA E Asia*

Small to medium-sized, up to 1 inch across. Flattened or slightly domed with a large, wide umbilicus which is strongly sculptured, often with prominent notches. Outer margin of last whorl sharp with a spiral ridge. Sculpture of a few spiral ridges above and below suture. Aperture sub-triangular and thickened at two outer angles. (*Stellaxis* can be confused with the related genus *Architectonica* whose whorl surfaces are covered with widely-spaced spiral grooves.)

Stellaxis (sundial shell)

"Aporrhais" *Cretaceous: Worldwide*

Modern range is restricted to North Atlantic. Large to medium-sized, up to 5 inches high. Turretted with long anterior spine or elongate canal. Outer lip of aperture flared and notched with a variable number of spines developed. Sculpture of strong axial ridges and tubercles; spiral sculpture also developed with ribs often growing into the spines of the outer lip.

Crepidula

2 ins

"Aporrhais"

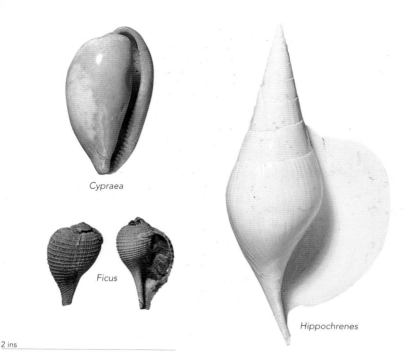

Cypraea

Ficus

Hippochrenes

2 ins

Cypraea (cowry) *Tertiary–Recent: Worldwide in warm waters*
Small to large, 0.25 to 6 inches long. Highly characteristic egg-shape with outer lip of last whorl greatly expanded and completely covering the rest of the shell. Aperture elongate, continuing along entire length of shell, with serrated margins and outer lip thickened. Surface usually smooth and shiny.

Ficus (fig shell) *Paleocene–Recent: NA E Asia*
Small to large, usually 0.5 to 5 inches long. Low spired, shell spindle-shaped (fusiform) with very large last whorl and lower region produced as broad, twisted canal. Whorls of spine rounded or with shoulders. Aperture large broad and elongate. Shell thin with little columellar callus. Sculpture of spiral and axial ribs.

Hippochrenes *Eocene: E Asia*
Medium-sized to large with tall spire, approximately equal to height of last whorl. Lower part of last whorl produced as elongate canal. Outer lip expanded as a large flare which is fused to the spire. The lower face of this carries a deep groove along the junction with the spire, and the shell below this groove may be expanded to cover it. The specimen shown has moderate development of the flare, but in some forms it may resemble that of *Aporrhais*.

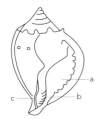

Galeodea: showing features

Volutospina: aperture region

Galeodea *Eocene: NA E Asia*
Medium-sized, usually 1 to 3 inches long. Like *Ficus* but with higher, conical spire. Whorls angular with strong spiny projections at shoulder; spiral ridges and more swellings of variable strength on last whorl. Aperture (a) elongate, with thickened outer margin carrying serrations. Strong columellar callus (c) with several strong ridges on inner margin (b), especially at lower end.

Volutospina *Paleogene: NA E Af Asia*
Medium-sized, usually 1 to 4 inches long. Representative of a group in which the spire is of intermediate height. Whorls angular usually carrying ribs which bear spines at the shoulder. Aperture (a) narrow with short canal (b). Columellar plications twisted (c).

Marginella *Eocene–Recent: Worldwide*
Small to medium-sized, less than 0.25 to 2 inches long. Often tapering equally at each end, oval or elongate. Surface smooth, unsculptured. Aperture (a) elongate (shown here), outer margin thickened and sometimes bearing teeth (not shown here). Several columellar plications present (b,c,d). Sutures slightly impressed.

Clavilithes *Eocene–Pliocene, ?Recent: NA E Asia*
Medium-sized to large, usually 4 to 6 inches long. Shell elongate, conical with sutures deeply impressed. Spire short, pointed and often strongly sculptured at the apex; lower parts of shell smooth. Broad almost flat ramp above shoulder; rest of whorl almost vertical. Whorls increasing uniformly in size. Long canal. Aperture as shown *(also see page 236)*. No columellar plications. A longitudinal section of the shell shows the ramp, whorl shape, canal, aperture, and columella.

2 ins

Galeodea

Volutospina

Marginella

Buccinum

244

2 ins

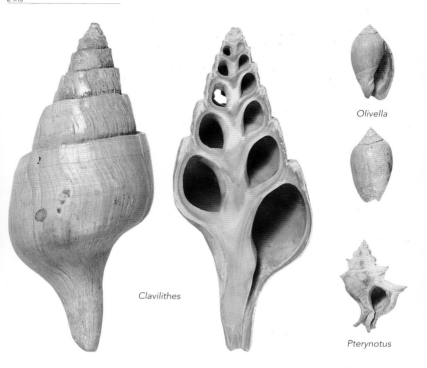

Clavilithes

Olivella

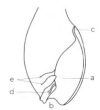

Pterynotus

Pterynotus *Eocene–Recent: Worldwide*
Usually 1 to 3 inches long. Dominant ornament of three strong axial ribs per whorl, each bearing a spine and with one smaller tubercle between each, overriden by spiral ridges. Aperture small. Outer lip expanded into rib as shown and having ridged inner margin. Inner lip thickened. Canal medium to long. Columellar plications absent.

Buccinum (whelk) *Pliocene–Recent: NA E*
Medium-sized to large, usually 1 to 6 inches high. Fusiform shell with whorls increasing uniformly in size. Aperture wide and oval with short canal. Ornament of spiral and/or axial ribbing. Outer lip sharp, sometimes recurved. Columellar callus relatively weak.

Olivella *Tertiary–Recent: Worldwide*
Usually less than 2 inches long. Last whorl very high in relation to rest of shell. Low spire. Aperture (a) elongate with short, wide canal (b) and notch at upper end (c). Columellar plications present (d). Outer lip thin and sharp. Very weak axial grooves may be present and several spiral grooves are usually developed near the lower end of the last whorl (e).

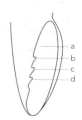

Marginella: aperture region

Olivella: aperture region

245

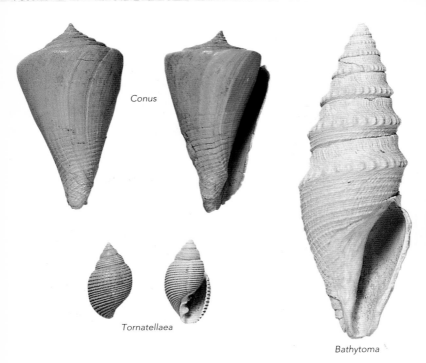

Conus

Tornatellaea

Bathytoma

2 ins

Tornatellaea: aperture region

Trochactaeon: aperture region

Conus *Eocene–Recent: Worldwide*
Small to large, usually 1 to 4 inches long. Steep upturned, conical shape below, with flat to steep conical spire above. Aperture parallel-sided, long and narrow (shown here) with a notch at the upper end. Canal short and outer lip thin. Ornament of spiral grooves, ridges or tubercles. Spiral ridges of variable strength on spire.

Bathytoma *Tertiary–Recent: Worldwide*
Small to medium-sized, usually 0.5 to 3 inches long. Shell narrow, equally conical at both ends. Last whorl about half total height. Aperture elongate, almost parallel-sided. Columellar plications absent. Sutures deep. Growth lines flexed backward at shoulder with row of strong tubercles along shoulder. Spiral ribs present.

Tornatellaea *Jurassic–Oligocene: Worldwide*
Small, usually less than 0.25 to 1 inch long. Spire whorls rounded. Aperture elongate, almost parallel-sided. Columellar plications absent. Sutures deep. Growth lines flexed backward at shoulder with row of strong tubercles along shoulder. Spiral ribs present.

Planorbis

Trochactaeon

2 ins

Trochactaeon *Cretaceous: Worldwide*

Large to medium-sized, usually 1 to 3 inches long. Spire low, concavely pointed, body whorl large. Aperture elongate, parallel-sided. Columella with two or three strong folds at the lower end. Shell smooth and thick. In a closely-related form *Actaeonella* (Cretaceous: NA) the last whorl is expanded and covers the spire, giving a superficial appearance similar to that of *Cypraea*.

Planorbis *Oligocene–Recent: E Af Asia*

Small to medium-sized, usually less than 0.25 to 2 inches long. Flattened spiral (planispiral) shell with upper side flattened (shown here) and lower side concave (a), or with broad umbilicus and sutures deeply impressed (shown here). A low spire is sometimes present (b), or both surfaces may be concave (c). Aperture oval to wide crescentic. Outer margin sharp. Ornament of fine growth lines only.

Planorbis: cross-sections showing grown forms

Cephalopods

Squid, octopus, cuttlefish, and nautiluses are living cephalopods: mollusks which have adopted an active, predatory lifestyle and typically reach sizes far larger than any other invertebrates (the modern giant squid *Architeuthis* may weigh up to 2,200 lbs). Cephalopods are characterized by a hollow shell divided up into chambers by septa. As the animal grows, these chambers are emptied, reducing the density of the animal, allowing it to become neutrally buoyant. The nautiluses and ammonites are common fossils in the Paleozoic and the Mesozoic, and are widely used for dating rocks (*biostratigraphy*).

Nautiluses

The most primitive cephalopods, possessing external shells either straight ("orthocones") or coiled, of which two genera (*Nautilus* and *Allonautilus*) are still alive today. In most nautiluses the septa are simple disks and quite widely spaced. Most diverse in the Paleozoic, the modern species are relatively deep-water scavengers which stay close to the sea floor.

Orthoceras *Ordovician: Worldwide*
Straight shelled forms with narrow siphuncles running along the center of the shell. Cameral deposits made of calcium carbonate are found in the apical part of the shell. These acted as a counterweight for the body of the animal at the front of the shell and allowed the nautilus to swim horizontally. Mostly of moderate size (less than 3 feet) but some as much as 10 feet in length.

Orthoceras

2 ins

Eutrephoceras

Aturia

Endoceras

Endoceras *Ordovician: NA E Asia Australia*
Straight shelled forms with very broad siphuncles usually displaced ventrally. Some species are extremely large, in excess of 30 feet in length.

Eutrephoceras *Jurassic–Miocene: Worldwide*
A wide-ranging, compact, medium-sized nautilus with a coiled shell ornamented with short ribs emerging from the umbilicus but otherwise smooth.

Aturia *Paleocene–Miocene: Worldwide*
Completely smooth and strongly laterally compressed, this nautiloid is characterized by closely-packed, fluted septa similar to those of the goniatites (see under Ammonites).

Ammonites

Not known after the Cretaceous. Ammonites are one of the most important groups of fossils for dating late Paleozoic and Mesozoic rocks, because they changed very rapidly with time and had wide geographical distributions.

Ammonite shells are externally similar to some gastropod shells, but their shells are divided by *septa* and they have *suture-lines* and a *siphuncle*. Most ammonites consist of *whorls* coiled in a plane spiral, but forms coiled in a helical spiral, as well as straight and hook-shaped forms, also occur. Whorls are often ribbed, and sometimes bear spines which are generally broken off, leaving only a series of low rounded bases *(tubercles)*. The *suture-line* is the junction between a *septum* and the outer shell, and is visible as a wavy, frilled line on the surface of the shell. Suture-line diagrams trace the line of the suture from the *venter* (a) to the *umbilical seam* (h), where adjacent whorls join. The arrow on a suture-line diagram points toward the *aperture* (k); *lobes* are the backward folds of the suture-line and *saddles* are the forward projections. A simple suture-line is shown here at the junction of the two colors on *Ceratites*, and a complex suture-line is shown on *Phylloceras* on page 251. The *siphuncle* (b) is a tube running through the septa and chambers near the venter of each whorl, and is often indicated by sharp flexions of the suture-lines. The *umbilicus* (d) is the depression on the side of the shell produced by the coiling, and the increasing size of the whorls, and the *umbilical shoulder* is where the shell turns inward (g) to the seam from the *lateral surface* (c). The *keel* (f) is a ridge which may be present along the venter. The thickness is (e) and arrow (j) points backward. Most specimens shown here are internal molds but the shell or *conch* is clearly shown on the lower part of *Goniatites*. Suture-lines are important for identification in some groups of ammonites; typically, early ammonites have simpler suture lines (like *Goniatites* and *Ceratites* here), and later forms have more complex sutures.

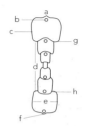

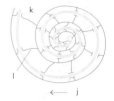

Typical structure of an ammonite: cross-section (top) and longitudinal section

Ammonitic structure showing saddles (s) and lobes (l)

Ceratitic structure showing saddles (s) and lobes (l)

2 ins

Gastrioceras

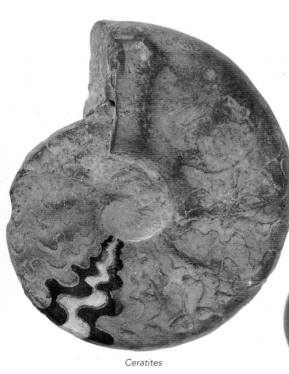

Ceratites

Goniatites

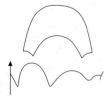

Goniatites: cross-section of whorl (top) and suture diagram

Paleozoic ammonites

Goniatites *Lower Carboniferous: Northern Hemisphere*
Thick, globular shells with narrow umbilicus. Suture-line is characteristic of the group (Goniatitida) which ranges through the Carboniferous and Permian. In this type the ventral lobe is pointed and the lateral lobe is smooth and convex. Part of the shell is shown here and carries fine growth lines.

Gastrioceras *Upper Carboniferous: Worldwide*
Globular whorls with narrow umbilicus (shown here), but some forms are flattened with wider umbilicus. Suture-line simple with smoothly curved saddles and pointed lobes. Ornamented with strong ribs near the umbilicus, which divide on the side of the whorl then extend as fine ribs across venter, flexing slightly backward in the center.

Gastrioceras: cross-section of whorl

Mesozoic ammonites

Ceratites *Triassic: E*

Moderately evolute whorls and wide umbilicus. Shape of suture-line is known as *ceratitic* and is characteristic of the group (Ceratitida) which ranges through the Triassic. Has smooth, simple saddles and serrated lobes. Forms similar to *Ceratites* are known from the Triassic of North America.

Lytoceras *?Upper Triassic, Lower Jurassic–Upper Cretaceous: Worldwide*

Evolute whorls only just in contact, wide umbilicus, and almost circular whorl cross-section. Whorls increase rapidly in diameter in some forms. Suture-line complex with ventral lobe divided, two lateral lobes and edges of all lobes extremely subdivided and fern-like. Ornamented with fine ribs. If the test is preserved, characteristic frilled ribs are present at intervals (not shown here).

Ceratites: cross-section of whorl

Phylloceras *Upper Triassic–Upper Cretaceous: Worldwide*

Involute, flattened whorls with umbilicus very narrow or closed and last whorl covering most of the earlier ones. Suture-line complex and shown at junction of red and white paint on specimen shown; saddles have characteristic leaf-like projections and lobes have pointed projections. Surface smooth or ornamented with fine striae extending unbroken across the venter.

Phylloceras: cross-section of whorl

Lytoceras

2 ins

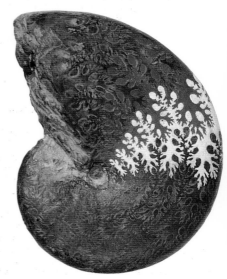

Phylloceras

251

Arnioceras

Asteroceras

Amaltheus

Dactylioceras

Promicroceras

2 ins

Arnioceras: cross-section of whorl

Jurassic ammonites Early Lower Jurassic

Arnioceras NA SA E Af Asia
Evolute whorls with very wide umbilicus. Ornamented with strong, simple ribs which bend forward near venter. Keel strong and bordered by furrows on each side.

Asteroceras NA E Asia
Medium-sized. Moderately evolute, thick whorls that increase in size rapidly, with wide umbilicus and keel on venter bordered by shallow depressions. Ornamented with strong ribs that fade on the venter. Suture relatively simple (shown here).

Promicroceras E
Small, up to 2 inches in diameter. Very evolute whorls, with circular cross-section, wide umbilicus and rounded, unkeeled venter. Ornamented with strong ribs sometimes flattened where they cross the venter.

Jurassic ammonites Middle Lower Jurassic

Amaltheus *Northern hemisphere*
Involute, flattened whorls, with fairly narrow umbilicus and strong
keel on venter. Ornamented with strong ribs that bend forward
and fade on outer part of whorl, then strengthen again to form
the characteristic serrations on the keel.

*Amaltheus: cross-section
of whorl*

Jurassic ammonites Late Lower Jurassic

Dactylioceras *Worldwide*
Evolute whorls with almost circular cross-section and wide umbilicus. Ornamented with sharp ribs, which divide on upper part of
whorl and are continuous across the venter. Tubercles occur on
the ribs at the point of division in some forms.

Harpoceras *NA SA E Af Asia*
Involute, flattened whorls, with small to medium-sized umbilicus
and strong ventral keel. Ornamented with strong flexuous or
sickle-shaped ribs, which bend backward in the middle of the
whorl then swing forward on the outer part and extend on to the
venter.

*Harpoceras: cross-section
of whorl*

Hildoceras *E Af Asia*
Evolute whorls with a quadrate whorl section and a wide umbilicus. Characteristically grooved near middle of whorl side. Ribs
absent or feeble on umbilical side of groove, but strong back-
wardly curved ribs on ventral side of groove. Strong ventral keel
bordered by deep furrows.

*Hildoceras: cross-section
of whorl*

2 ins

Harpoceras

Hildoceras

Jurassic ammonites Middle Jurassic

Parkinsonia *E Af Asia*

Evolute whorls, with medium to wide umbilicus. Ornamented with sharp ribs that bifurcate and bend forward near the edge of the venter, then end at the edge of a deep groove in the middle of the venter which becomes shallower at larger sizes.

Stephanoceras *Worldwide*

Evolute whorls with almost circular cross-section and wide umbilicus. Strong ribs on the inner part of the side of the whorl divide into two or three at a mid-lateral tubercle, then continue across the venter with a mid-ventral interruption.

Graphoceras *E Af Asia*

Involute flattened whorls with a medium to small umbilicus and a strong keel on the venter. Ornamented with flexuous ribs similar to those of *Harpoceras*, which diminish in strength at larger sizes. The specimen shown here is complete with most of its shell, and growth lines following the shape of the ribs are visible near the aperture.

Graphoceras: cross-section of whorl

Jurassic ammonites Upper Jurassic

Cardioceras *NA E Asia*

Moderately involute whorls with triangular cross-section, moderate to narrow umbilicus and keel on venter. Strong ribs divide into several smaller ribs at middle of whorl side, then curve forward to form forwardly pointing chevrons (serrations) on the keel. Ribs are fine on early whorls, becoming more widely spaced at larger sizes. (The specimen shown here is the subgenus *Scarburgiceras*, with less distinct keel and less differentiated ribs.)

Cardioceras: cross-section of whorl

Perisphinctes *E Af Asia*

Some specimens attain very large sizes. Very evolute whorls with almost square cross-section and steep umbilical wall. Strong, straight ribs divide into two or three near edge of venter, then cross the venter without interruption. Ribs on side of whorl suddenly become very large, bold and single at large sizes, fading on venter at the same time.

Pavlovia *E Asia*

Evolute whorls with whorl cross-section broadest near edge of umbilicus. Sharp, wiry ribs divide into two near middle of side of whorl and are continuous across venter.

Pavlovia: cross-section of whorl

2 ins

Pavlovia

Graphoceras

Parkinsonia

Stephanoceras

Perisphinctes

Cardioceras (Scarburgiceras)

Cretaceous ammonites Lower Cretaceous

Hamites: outline of complete shell

Hoplites: cross-section of whorl

Hamites Worldwide
Uncoiled with characteristic shape as shown, usually with straight shell between two or three hooks, but some forms have helical coiling. Many specimens are fragmentary, and only the shaded portion is shown here. The whorl has a circular or oval cross-section, and is ornamented with ribs that entirely encircle the shell. Suture-line simple.

Hoplites E Asia
Medium-sized, involute whorls, with deep, narrow umbilicus. Tubercles at edge of umbilicus give rise to strong ribs that quickly divide and curve forward at the edge of the venter, which has a smooth central depression.

Mortoniceras NA SA E Af Asia
Large, flattened, evolute whorls, with a wide umbilicus, quadrangular whorl section and a strong keel bordered by furrows on venter. Ornamented with strong ribs that bear tubercles near edge of umbilicus. Most ribs divide into two on side of the whorl then curve forward up to the edge of the venter. Ribs are often serrated on outer part of whorl side.

Douvilleiceras NA SA E Af Asia
Globular, swollen whorls, evenly rounded in cross-section, with medium to narrow umbilicus. Strong, single ribs bear many tubercles on side of whorl and venter, but are interrupted along the mid-line of venter.

2 ins

Hamites

Hoplites

Douvilleiceras

Mortoniceras

2 ins

Oxytropidoceras

Scaphites

Placenticeras

Turrilites

Douvilleiceras: cross-section of whorl

Oxytropidoceras *NA SA E Af Asia*
Involute, flattened whorls with narrow, deep umbilicus and strong keel on venter. Ornamented with numerous flexuous ribs curving forward at their ventral ends.

Oxytropidoceras: cross-section of whorl

Placenticeras *NA E Af*
Involute, flattened whorls with narrow, deep umbilicus and raised, flattened *(tabulate)* venter. Ribs weak or absent, but pointed tubercles border the umbilical edge; another row of tubercles on outer part of whorl, and small tubercles on side of flat venter in some forms.

Cretaceous ammonites Lower–Upper Cretaceous

Turrilites *NA E Af Asia*
Coiled in a helical spiral like some gastropods (from which they are distinguished by possessing septa and suture-lines). Ornamented with coarse ribs and tubercles. Suture-line complex (not shown here).

Placenticeras: cross-section of whorl

Scaphites *Worldwide*
Normal spiral whorls are followed by a short straight section then a hook-shaped living chamber. Ornamented with many fine branching ribs and tubercles; the ribs curve forward and are continuous across the venter.

Scaphites: cross-section of whorl

Cretaceous ammonites Upper Cretaceous

Baculites *Worldwide*

Up to 80 inches long. Only the very earliest (smallest) part of the shell is coiled and is rarely seen in collections; remainder of the shell is a single, long, straight shaft, usually found as fragments. Cross-section flattened or oval, and suture-line complex. Shell smooth or ornamented with sinuous ribs.

Baculites: cross-section of whorl

Acanthoceras *NA E Af Asia*

Evolute, robust whorls with wide umbilicus, quadrate whorl cross-section and keel on venter. Ornamented with straight ribs bearing tubercles at edge of venter.

Acanthoceras: cross-section of whorl

Belemnites

An extinct cephalopod group, superficially like modern squid. Unlike ammonites, the shell was internal. The bullet-shaped back part of the shell, the *guard*, is dense and durable, and is a common fossil in the Jurassic and Cretaceous. A hollow region at the front of the guard, the *alveolus*, houses the cone-shaped chambered part of the shell, the *phragmocone* (shown here in *Cylindroteuthis*). In rarely preserved, complete specimens, a flattened region, the *pro-ostracum*, is an anterior projection of the phragmocone which lies over the mantle cavity. Broken specimens of the guard show a structure of radiating calcite fibres and concentric growth lines.

2 ins

Baculites   *Belemnitella*   *Acanthoceras*

Neohibolites *Upper Cretaceous: Worldwide*
Small, guard usually 2 to 4 inches long. Circular cross-section. From its widest central part, the guard narrows towards the phragmocone, which is preserved but crushed in the specimen shown, and seen at the front end. A short slit and groove are present on the guard at the back, near the alveolus, as shown here.

Belemnitella *Cretaceous: NA E Asia*
Large, guard usually over 4 inches long. Cross-section almost circular but has a flattened upper surface with a pair of shallow, longitudinal depressions. There is a long slit on the ventral surface of the alveolus near its edge.

Cylindroteuthis *Jurassic–Early Cretaceous: NA E*
Large, guard up to 6 inches long. Cross-section almost circular with slightly flattened sides, and long groove on the ventral surface. The chambered phragmocone is clearly shown here at the front of the specimen.

Neohibolites

Scaphopods

A major group of mollusks of the same status (Class) as the gastropods, cephalopods, and bivalves, but less common in fossil and Recent faunas, and more uniform in appearance. Shells elongate, conical, open at both ends and curved like an elephant's tusk. In life the concave side is upward; the smaller (*apical*) opening is posterior, and the larger opening is anterior, this being buried in the sediment.

Forms with numerous ribs (e.g. *Prodentalium*, Carboniferous–Recent) acquired an apical slit during the early Tertiary (e.g. *Fissidentalium* shown here). This is the central stock of the ribbed scaphopods, and *Dentalium*, symmetrical and with few ribs, which is now dominant, did not appear until the Miocene.

Fissidentalium *Paleocene–Recent: Worldwide*
This genus has up to 40 unequal ribs asymmetrically arranged around the tube. The underside of the apex usually has a long slit, which is absent from the earlier *Prodentalium*. (The genus *Dentalium* Miocene–Recent, has 6–16 primary ribs symmetrically placed, and an apical notch on the underside).

Fissidentalium

Cylindroteuthis

2 ins

259

Bivalves

Cockles, scallops, razor shells, oysters, mussels, and clams are all bivalve mollusks. Bivalve mollusks superficially resemble brachiopods *(page 280)* but closer inspection reveals important differences. In most bivalve mollusks each valve is asymmetrical *(inequilateral)* with the beak toward the front end; and the valves are mirror images of each other *(equivalve)*. Oysters are well-known exceptions to this and they are *inequivalve*. In brachiopods each valve is usually symmetrical but the two valves differ in size and curvature.

Important features of bivalve mollusks are height (h), length (l), thickness (t), the beak (b), and ornamentation. A flattened region between the beaks of paired valves is an *area (shown in Arca on page 265)*; a flattened depression in front of the beak is a *lunule* and behind the beak is an *escutcheon*. An opening or notch between or behind the beaks is a *ligamental notch*. Valves articulate at the "hinge line" *(dorsal margin)*. In some shells the valves do not meet at the front or back and a *gape* is left *(shown in Pholadomya on page 262)*.

Inward projections of shell known as *hinge teeth*, collectively referred to as *dentition*, may be present below the beak and at either end of the dorsal margin; ridges or "teeth" on the side and lower margins are termed *crenulations (shown in Glycymeris on page 265)*. Front and back muscle scars may be present (shown here in *Mya*) or only one may be developed. The *pallial line* is a curving linear mark joining the front and back muscle scars indicating the extent of attachment of the animal within the shell. The *pallial sinus* is an inflexion of this line near the back of the shell. The shell exterior is often sculptured ("ornament") with features such as *concentric ribs, radial ridges,* and *spines,* according to mode of life.

Typical features of a bivalve shell: side view (top) and cross-section

2 ins

Venericor

Arctica

Venericor *Paleocene–Eocene: NA SA E Af*

Ranging from 1 to 6 inches long. Equivalve and strongly convex. Strong beak points forwards. Ligamental notch behind beak. Two strong teeth (a,b) under beak on each valve as shown. Ornament of wide radiating ridges and concentric lamellae strongest near margin. Margins with small crenulations.

Venericor: showing teeth

Arctica *Paleocene–Recent NA E*

Usually 1 to 4 inches long. Shape similar to *Venericor*. Ligamental notch deep. Two or three teeth present as shown. Pallial sinus absent. Ornament of weak concentric ridges. Margins lacking crenulations.

Orthocardium *Paleocene–Eocene: E*

Representative of group which includes cockles. Medium-sized. Valves almost symmetrical; beak points slightly forward. Dorsal margin almost straight. Two central teeth on each valve, one side tooth at front and back on left valve; two front and one back on right valve. Ornament of strong frilly ribs with beaded edges. Margins with strong crenulations.

Arctica: beak showing teeth

Mya *Oligocene–Recent: NA E Asia*

Usually 1 to 6 inches long. Elongate, flattened. Beak small, pointing upwards. Ornament of concentric lamellae or smooth. Dorsal margin curved, lacking teeth but having a spoon-like process known as the *chondrophore*. Crenulations absent. Wide posterior gape. Front muscle scar high and curved; back muscle scar circular and deep. Deep pallial sinus.

Mya

Orthocardium

2 ins

Teredo

Pitar

Neocrassina

2 ins

Pitar: beak and hinge
showing teeth

Teredo *Eocene–Recent: Worldwide*

Representative of a group of mollusks most commonly known from their borings in wood (shown here). Burrows are circular in cross-section and may have a calcareous lining. They may be filled with mud or contain remains of the shell. *Teredo* has a very small shell and the grouping of Recent genera in this group is based on the soft anatomy.

Pitar *Eocene–Recent: Worldwide*

Medium-sized. Valves very convex and similar in shape to *Arctica*. Beak points forward, lunule shallow, escutcheon absent. Teeth as shown; front, side teeth well developed, usually three central teeth (a) in each valve. Ligamental notch behind beak, otherwise margins closed; lower margin smooth. Pallial sinus present. Ornament of weak concentric ridges.

Neocrassina *Jurassic–Cretaceous: E Af*

Medium-sized; shallowly convex to thick. Beak points forward and front part of shell much smaller than back. Large lunule and escutcheon clearly defined. Ornament of concentric ridges. Two central teeth on each valve. Margins smooth in younger specimens developing small crenulations later in life, perhaps reflecting change from male to female. Margins closed.

Pholadomya *Triassic–Recent: Worldwide*

Medium-sized to large, elongate. Valves very convex, shell thin.

Beak near front end, not strong, rounded and pointing upward. Ornament of radiating ridges over central region but with concentric ridges prominent at front and back ends. Teeth absent or weak. Valves with strong back gape and smaller front gape. Pallial sinus present.

Sanguinolites Devonian–Permian: Worldwide
Medium-sized. Elongate and curved with front end very reduced. Thick. Teeth absent from dorsal margin; escutcheon large and clearly defined; lunule less well defined. Ornament of concentric ribs. Margins smooth and leaving small gape at back end.

Trigonia Triassic–Cretaceous: Worldwide
Left valve shown. Medium-sized to large. Almost triangular with front edge steeper than back. Beak pointing upward or slightly backward. Flattened face at back of shell delimited by high ridge and smooth channel. Ornament at front of strong concentric ridges and at back of weaker radiating ridges. Escutcheon large and defined by a high crest with a beaded edge. Large central tooth (c) on left valve, and two large teeth (d) on right valve, have strongly grooved surfaces. Margins closed and smooth.

Trigonia: beak showing teeth on left (top) and right valves

Schizodus Carboniferous–Permian: Worldwide
Small to medium-sized. Thick with flattened margins. Beak strong, pointing upward and front end reduced. Lunule and escutcheon absent. Single large tooth on each valve (a), a few smaller teeth also present. Shell surface smooth or with weak concentric ripples. Margins smooth and closed.

Schizodus: beak and hinge showing tooth

2 ins

Pholadomya

Trigonia

Schizodus

Sanguinolites

2 ins

Anodonta

Modiomorpha

Carbonicola

Anodonta (swan mussel)
Cretaceous–Recent (fresh water): NA SA E Af Asia
1 to 6 inches long. Shell elongate, beak well-formed pointing forward or upward. Shell flattened to thick. Surface smooth or with concentric rings. Hinge toothless or with small ridges. Ridge of variable strength runs backward from the beak to the back margin. Back margin more pointed than front. Margins closed and lacking crenulations.

Carbonicola *Carboniferous: E (non-marine)*
Fresh water, medium-sized, flattened to thick. Elongate at back, shortened at front. Beak pointing upward or forward. Dorsal margin curved. Sometimes one or two tooth-like structures present under beak on each valve. Margins smooth, closed. Front muscle scar circular and deep; back scar shallow and high. Ornament of concentric lines.

Modiomorpha *Silurian–Devonian: NA E Asia*
Medium-sized. Equivalve and valves expanded backward. Beak low. Single tooth on left valve and socket on right. Margins smooth and closed. Ornament of concentric lines.

Arca *Tertiary–Recent: Worldwide*
Medium-sized, usually 2 to 4 inches long. Elongate with beak well in front of mid-line and pointing slightly forward. Valves very convex. Dorsal margin carrying very wide, flattened areas which separate the beaks. Hinge with long row of small, comb-like teeth. Lower margin with elongate gape between the valves. Ornament of concentric and radial ribs.

Parallelodon *Devonian–Jurassic: Worldwide*
Usually 2 to 6 inches long. Elongate with very long back region and shortened front end. Beak pointing forwards. Dorsal margin straight. Large flattened areas between beaks carry longitudinal ridges. Very few teeth near back of hinge (a), and numerous shorter, curving teeth near front (b). Elongate gape on lower margin. Margins smooth.

Parallelodon: beak and hinge showing teeth

Glycymeris *Cretaceous–Recent: Worldwide*
Small to medium-sized, almost circular. Beak almost centrally placed and pointing upwards (equilateral). Teeth like *Arca* but arranged in gentle curve. Areas developed but smaller than in *Arca*. Crenulations on lower margin. Surface smooth or with radial ridges and concentric grooves.

Modiolus *Triassic–Recent: Worldwide*
Medium-sized to large, up to 4 inches long. Generally similar to the common mussel *Mytilus* but beak not at very front of shell. Dorsal margin without teeth. Shell surface smooth or with shallow concentric ridges. Equivalve with ligamental notch developed, otherwise margins smooth and closed.

2 ins

Dorsal (umbonal) view of left valve of *Arca*

Arca

Parallelodon

Glycymeris

Modiolus

265

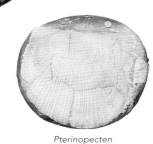

Atrina (pen shell)

Pterinopecten

Inoceramus

2 ins

Gervillella

Inoceramus: beak and hinge
showing ligamental pits

Atrina (pen shell) *Eocene: Worldwide*
Medium-sized to large. Shaped like a half-closed fan, triangular
and up to 10 inches long. Valves equal with beaks at anterior
point. Ornament of wide ripples below and radiating ridges
above. Shiny inner shell layer surface often exposed (shown here).
Lower margins with elongate gape near front and back margins
wide open. (*Pinna* Jurassic–Recent: Worldwide, is a similar and
related genus which is more elongate and differs internally from
Atrina.)

Gervillella *Triassic–Cretaceous: Worldwide*
Medium-sized to large, up to 10 inches long. Very elongate with
greatly lengthened back and reduced, sharply pointed front.
Dentition of a few elongate teeth which are almost parallel to the
long axis. Region above dorsal margin flattened with numerous
(up to ten) vertical pits which hold the ligament. Ornament of
concentric lamellae.

Inoceramus *Jurassic–Cretaceous: Worldwide*
Medium-sized to large, usually 3 to 6 inches high. Back wing
expanded as shown or reduced. Numerous ligamental pits (a)
along upper edge of dorsal margin and wing. Hinge without
teeth. Ornament of concentric, coarse ripples and fine grooves.
Beak points upward. Shell short and high, very convex.

Pterinopecten *Silurian–Carboniferous: Worldwide*
Medium-sized. Beak pointing upward. Dorsal margin straight
with wings developed before and behind beak; back wing larger.
Right valve usually less convex than left. Ornament of radial
ridges of variable strength.

2 ins

Oxytoma

Meleagrinella

Oxytoma *Triassic–Paleocene: Worldwide*
Small to medium-sized. Beak pointing upward with wings developed before and behind. Back wing usually longer and pointed.
Right valve flattened, left valve convex. Hinge lacking teeth but
with narrow areas, that of the left valve continuing in plane of
margin and that of the right valve is at about 90° to this. Ornament of coarse ridges with wide intervals. Ridges produced as
spines around margin.

Meleagrinella *Triassic–Jurassic: Worldwide*
Small to medium-sized. Small wings before and behind beak.
Hinge lacking teeth. Left valve convex, right valve flattened. Left
valve with radial ridges which have spiny edges; ridges weak or
absent on right valve. A block with many small specimens is
shown with mainly left valves visible.

Chlamys (scallop) *Triassic–Recent: Worldwide*
Medium-sized, rarely more than 6 inches high. Similar to living,
common scallops. Equilateral, inequivalve, left valve more convex
than right. Wings before and behind beak; front wing notched on
left valve. Hinge teeth absent but triangular ligamental notch
developed under center of beak on both valves. Ornament of
strong ribs giving serrated edges at the margins. Concentric
ornament is also usually developed. (*Chlamys* is one of several
genera referred to loosely as "scallops".)

Chlamys

Gryphaea *Triassic–Jurassic: Worldwide*
Medium-sized to large, up to 6 inches long. Left valve much larger than right and very convex with beak rolled over onto right valve and displaced slightly backward. Right valve flat or concave. Ornament of left valve numerous well-defined lamellae. Right valve with smooth or rippled surface and lamellae near margin. Left valve with elongate curved swelling along back edge above margin.

Actinostreon *Jurassic: Worldwide*
Usually medium-sized. Valves convex, shape varying from similar to *Ostrea* to inequilateral (shown here). Almost equivalve. Radial ridges characteristic, varying from strong ripples to high ridges (shown here); these give lower margin zigzag contact. Inner faces with small tubercles near margins.

Ostrea *(common oyster) Paleocene–Recent: Worldwide*
Medium-sized to large, up to 8 inches long. Left valve moderately convex; right valve slightly smaller than left and flat. Shape varies from circular to more height than length. Left valve with irregular rounded ribs, crossing lamellae. Right valve unribbed with lamellae.

2 ins

Gryphaea

Cardiola

Ostrea

Actinostreon

Spondylus *Jurassic–Recent: Worldwide*
Medium-sized to large, up to 6 inches high. Nearly equilateral, strongly inequivalve. Valves high and right valve deeper than left. Dorsal margin straight. Beak of right valve with large area (c) which carries fine vertical and cross-striations. Area (f) of left valve low and sloping outward. The specimen shown here is particularly spiny but in some forms the ridges predominate with only a few spines present. Two large teeth (e) are far apart on the left valve and close together on the right valve. Deep notch (d) below center of beak on both valves.

Plagiostoma *Triassic–Cretaceous: Worldwide*
Medium-sized to large, up to 6 inches long. Valves same size. Beak points upward and the front edge is straight with an elongate, wide lunule. Margins usually closed. Teeth weak or absent. Surface smooth with fine concentric or radial striations.

Cardiola *Silurian–Devonian: NA E*
Small, beak points upward or forward, equivalve. Hinge teeth obscure. Triangular areas on both valves. Margins may have a gape. Strong radial ribs crossed by concentric grooves give a squared pattern. A block with external molds and impressions is shown here.

Nucula *Cretaceous–Recent: Worldwide*
Small, equivalve, beak points backward. Comb-like teeth along margins before and behind beak (shown here). Internal ligamental process under beak. Lower margin has fine striations. Anterior and posterior muscle scars equal in size. Outer surface smooth with concentric rings. Inner surface shiny (shown here).

Spondylus: beak and hinge of left (top) and right valves

2 ins

Nucula

Spondylus

Plagiostoma

Arthropods

The largest phylum of animals; includes insects, spiders, scorpions, crustaceans, millipedes, centipedes and several extinct groups, of which the trilobites *(pages 270 to 277)* are the most important. The group was well established by the start of the Cambrian. The most characteristic feature of the group is the hard outer coating, which is slightly flexible in most arthropods and provides attachment for the muscles. In most arthropods the body is divided into a head, thorax, and abdomen, with the jointed legs attached to the thorax. With the exception of trilobites, the arthropods are relatively uncommon as fossils, though insects and crustaceans may be locally abundant. The arthropod groups are extremely large and it is possible to show only a few representatives of the phylum here.

Trilobites

The most common fossil arthropods. The body is divided transversely into the *head* (a), *thorax* (b), and *tail* (c). It is divided along its length by two furrows delimiting the central *axis* (d) from the side regions. The axial region of the head *(glabella)* (e), is flanked on each side by the *genae* or *genal regions* (f). On the thorax and tail the sides are termed *pleural lobes* (g).

2 ins _____

Dalmanites

Ogygopsis

Eyes may be present on either side of the glabella *(shown in Phacops on this page)*. The rear outer corner of each genal region is termed the *genal angle* (h) and may project as a *genal spine* (i) *(shown in Dalmanites on this page)*. A *front border* may be present; this is a raised rim around the front of the glabella and genae.

The thorax consists of *segments* defined by *thoracic grooves* (j) and the number of these is important. The side region of each segment is a *pleuron*. A *pleural furrow* (k) is a groove sometimes present on the upper face of each pleuron.

The tail also shows segmentation, and transverse furrows may be present on the axis. The tail is known technically as the *pygidium* but this term is not used here.

The undersurface of the trilobite is only rarely exposed, but a large plate, the *hypostome*, may be locally very common. This comes from the underside of the head, and probably covered the foregut, behind which was the mouth.

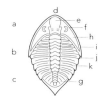

Typical structure of a trilobite as shown by Dalmanites

Dalmanites *Silurian: Worldwide*
Medium-sized. Tail about same size as head. Glabella with deep grooves, widening forward; eyes prominent; front border wide; genal spines long. Thorax about 11 segments; pleural furrows marked. Tail about 11 segments; back border smooth, carrying a spine. Ornament of small tubercles.

Phacops *Silurian–Devonian: Worldwide*
Head larger than tail. Glabella wide and widening forward; eyes large, lenses visible here; front border convex and bounded by deep groove; genal angles rounded. Thorax about 11 segments. Back edge rounded and smooth. The specimen shown here is rolled up.

Phacops

Ogygopsis *Cambrian: NA*
Medium-sized to large. Elongate; tail larger than head. Glabella parallel-sided with faint cross-grooves; eyes long and narrow; front border wide and flattened; genal spines short (not shown here). Thorax about eight segments; axis strong and wide; pleurae with deep wide pleural furrows. Tail about ten segments; axis tapering; tail pleurae with deep segmental grooves and furrows; back edge with convex border and smooth outline.

Calymene *Silurian–Devonian: Worldwide*
Medium-sized. Tail smaller than head. Glabella very convex and sloping steeply at the front, narrowing forward and carrying three pairs of swellings; eyes large; front border convex and separated from glabella by deep groove, genal angle rounded. Thorax about 13 segments. Tail about six segments.

Calymene

2 ins

Paradoxides   *Cambrian: NA E Af Aust*

Head much larger than tail. Glabella expanded forward, carrying about three pairs of cross-furrows; eyes large; genal spines about half body length. Thorax of about 18 segments; pleural furrows strong and diagonal; pleurae produced as spines at sides, which increase in size backward. Tail small with straight back edge.

Paedeumias   *Cambrian: NA E Asia*

Head large with flattened cheeks. Glabella deeply furrowed. Rounded swelling at front of glabella is connected to front border by a ridge; genal spines long. Thorax about 14 segments, decreasing in size backward from the second; first segment with short spine; second large with spine extending back beyond tail region; other pleurae with long spines; pleural furrows deep. Tail small, carrying long spine (twisted to left in the specimen shown here).

2 ins

Paradoxides

Paedeumias

Olenoides

2 ins

Encrinurus

Leonaspis

Cheirurus

Olenoides *Cambrian: NA SA Asia*
Head and tail about same size. Glabella with several furrows,
expanding slightly forward and reaching front border; eyes medium-sized; front border convex and wide; genal spines short. Thorax about seven segments; axis wide and tapering with
cross-furrows and tubercles or spines on each segment; pleurae
produced as short spines; Tail of at least five segments with axis
tapering backward; back edge with several pairs of spines.

Oryctocephalus *Cambrian: NA SA E Asia*
Tail and head almost equal in size. Glabella parallel-sided with
three or four pairs of cross-furrows which have deep pits at each
end; eyes small; genal spines long (not clearly shown here). Thorax about seven segments; pleurae produced as spines; pleural
furrows deep and diagonal. Tail axis with six cross-grooves; sides
and back of tail produced as long spines (not clearly shown here).

Oryctocephalus

Encrinurus *Silurian: Worldwide*
Head larger than tail. Glabella widening forward; eyes pronounced; genal spines small, directed outward. Thorax of 11 or 12 segments. Tail of five to ten pleural segments; back edge serrated.
Ornament of strong tubercles.

Cheirurus *Silurian–Devonian: Worldwide*
Tail smaller than head. Glabella produced forward to overhang
front border; eyes medium-sized; genal spines small. Thorax
about eleven segments; pleural furrows short and diagonal. Tail
with well-defined, deeply grooved axis; back edge with three
pairs of spines separated by small central spine.

Leonaspis *Silurian–Devonian: NA SA E*
Head very wide; eyes large; front border with strong spines; genal
spine large (broken off in the specimen shown here). Thorax
about eleven segments; pleurae produced backward as spines.
Tail small; back edge with one pair of large spines and two pairs
of smaller spines.

2 ins

Eodiscus (tail)

Eodiscus (head)

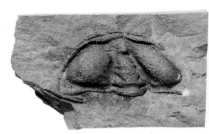

Triplagnostus

Ctenocephalus

Triplagnostus Cambrian: NA E Asia Aust

Small, less than 0.5 of an inch. Head and tail same size. Glabella divided into triangular front and elongate hind lobes, less convex than in *Eodiscus*; cheeks curved and divided at front by a groove; front border strong and convex; eyes absent; genal angle rounded or with small genal spine. Thorax of two segments. Tail very similar to head; axis of tail slightly wider than glabella, divided into larger triangular back region and a shorter front region which may carry a strong swelling. A groove at the back separates the two curved side regions of the tail. Back border similar to front border.

Eodiscus Cambrian: NA E

Very small, less than 0.25 of an inch. Head and tail same size. Head consists of a short, very convex glabella carrying a single pair of indistinct furrows, and curved cheek regions which are divided at the front by a deep groove extending to the narrow front border; eyes absent; genal angle sharp or strong; genal spines may be present. Thorax of two or three segments. Tail axis pronounced and carrying many strong cross-grooves; tail pleurae swollen and curved. Head and tail regions are here shown separated, and *Eodiscus* is often found in this condition.

Cedaria Cambrian: NA

Head and tail almost equal in size. Glabella lacking furrows, having rounded front end and terminating well behind border; eyes medium-sized; front border strong and convex; genal spines (not shown here) fairly long. Thorax about seven segments; axis well

2 ins

Cedaria   *Elrathia*   *Bonnaspis*

defined by furrows; pleural furrows long. Tail with strong axis and four or five furrows; back edge rounded.

Ctenocephalus   *Cambrian: NA E Af Asia*
Only head region shown. Glabella very convex, tapering forward, carrying three pairs of strong furrows; cheeks swollen, convex; eyes absent; front border very convex; genal spines long, extending over half the length of the thorax (not shown here). Body similar in shape to that of *Elrathia* with small tail and about 15 thoracic segments. Ornament of fine tubercles covering head region.

Bonnaspis   *Cambrian: NA*
Head slightly larger than tail. Glabella very convex, expanding strongly forward to front edge; furrows not present on glabella; eyes small; genal spines short (not shown here). Thorax about seven segments; pleurae with deep furrows. Tail up to five segments, poorly defined; back edge rounded.

Elrathia   *Cambrian: NA*
Medium-sized. Head much larger than tail. Glabella tapering forward with rounded front end well behind front border; glabella surface carrying several pairs of weak furrows; eye ridges strong; front border wide; genal spines short. Thorax about 13 segments; pleural furrows long and deep. Shallow furrows on tail indicate about five segments, back edge smoothly rounded. The most frequently encountered North American trilobite.

2 ins

Cryptolithus

Trinucleus

Cryptolithus *Ordovician: NA E*
Head much larger than tail. Glabella narrow and very convex, widening forward and carrying a single pair of furrows; eyes not visible. The most characteristic feature is the wide front border which slopes downward and outward, and carries radiating rows of deep pits. Genal spines long. Thorax about six segments. Tail smooth with raised central region and smooth back edge.

Trinucleus *Ordovician E*
Similar in general shape to *Cryptolithus*. Glabella convex and carrying three pairs of deep furrows; front border wide, carrying radiating grooves; genal spines long (not shown here). Thorax of six segments; axis strong. Tail much wider than long; back edge smooth.

Bumastus *Ordovician–Silurian: Worldwide*
Elongate with head and tail regions equal in size. Glabella not clearly defined but head carries large swellings on either side; genal angles rounded. Thorax of eight to ten segments; axis not clearly defined. Tail convex with steep back border and smooth outline. Surface ornament very weak.

Harpes *Devonian: E Af*
The head region only is shown here. Glabella very convex with lobes at sides; eyes strong; genal spines almost as long as body and very wide; front border wide, carrying many fine pits and tubercles. Thorax about 29 segments. Tail small.

2 ins

Bumastus

Isotelus

Griffithides *Carboniferous: NA E*
Medium-sized; elongate. Head and tail almost equal in size. Glabella wide and expanding slightly forward; eyes small; front border narrow; genal angle rounded. Thorax about nine segments; axis strong. Tail of numerous segments.

Isotelus *Ordovician: NA E Asia*
Head and tail equal in size. Glabella not clearly defined; eyes medium-sized, produced as conical swellings; genal angles rounded. Thorax of eight segments; axis very wide and defined by shallow furrows. Pleurae short; pleural furrows short, deep, and diagonal. Tail region pointed with weakly defined axial region and weak furrows in the pleural areas.

Harpes

Griffithides

Eurypterids

Eurypterus *Ordovician–Carboniferous: NA E Asia*
An extinct group closely related to the scorpions and important during the Paleozoic. Some eurypterids attained great size, being well over 3 feet long. Complete specimens are rare but fragments may be locally common. Eurypterids are popularly known as giant water scorpions and the largest, *Pterygotus*, was about 10 feet long, and is also the largest known arthropod.

Crustaceans

Lobsters, crabs, crayfish, shrimps, prawns, barnacles. One of the most important and diverse groups of marine invertebrates.

Hoploparia *Cretaceous–Eocene: Worldwide*
A small lobster. Note the jointed legs, large *chelipeds* (pincers) and long, segmented abdomen.

Balanus (barnacles) *Eocene–Recent: Worldwide*
Highly specialized crustaceans, sedentary as adults, with rigid plates. Opercular valves, across the opening, are retained in this specimen.

Insects

Body divided into three parts: head, thorax, and abdomen. Thorax has three pairs of legs. Wings usually present.

Libellulium *Upper Jurassic – Lower Cretaceous: Europe*
Libellulium is a dragonfly and belongs to the Order Odonata. Large predatory insects which first appeared in the Upper Carboniferous. They have two pairs of equal-sized wings with a dark spot near the tip, a long abdomen, and a large head with large eyes and short antennae. True dragonflies have permanently outstretched wings with veins that form a distinctive triangle near the base. Fossil forms are mostly known from incomplete wings, though the Solnhofen Limestone of Germany is famous for its complete specimens, as shown here. Wing veins are numerous and are used to identify fossil species.

Snipe fly *(Diptera: Rhagionidae) in Baltic amber.*
Amber is fossilized tree resin which was produced by trees for defense. It sometimes contains insects which are perfectly preserved. Although it is hard to find amber it can be purchased from jewelry shops. Most commercial amber comes from the Baltic region (U. Eocene) or the Dominican Republic (L. Miocene). Flies (Order Diptera) are the most common insects in Baltic amber and are distinguished from other insects by having only one pair of wings.

2 ins _____

Eurypterus

Hoploparia

Balanus

Snipe fly

Libellulium

2 ins _____

Brachiopods

Generally similar in appearance to bivalved mollusks as they consist of two shells. Distinguishing features of mollusks are given on page 236.

Important features are the *hinge line* (a); *interarea* (b); the flattened regions often present between hinge line and *beak* (c); front end (anterior commisure) (d); the fold which is a long swelling (visible here on the dorsal valve of *Spirifer*); and the *sulcus* which is a long channel (visible here on the ventral valve of *Spirifer*). The fold and sulcus often occur together on opposite valves. The ornamentation usually consists of radiating ridges (*ribs*) (as in *Spirifer*); and the *sulcus* which is a long channel (visible here on the ventral valve of *Spirifer*). The fold and sulcus often occur together on opposite valves. The ornamentation usually consists of radiating ribs (as in *Spirifer*), but concentric growth lines may also be present (as in *Atrypa*). The two valves are termed the *ventral* and *dorsal* valve. The ventral valve (e) always has the stronger beak, and is often larger than the dorsal valve (f). Also, the beak of the ventral valve often carries a small hole, the *foramen* (g), through which the attachment stalk (*pedicle*) emerges in the living animal.

Traditionally, brachiopods have been separated into two groups, Inarticulata and Articulata, based on whether the two valves of the shell were articulated or not. However, the most recent classification groups them into three: the Linguliformea, Craniformea, and Rhynchonelliformea. The valves of the first two groups are inarticulated, and those of the latter are articulated. Division into "inarticulated" and "articulated" has therefore been retained here for convenience.

Typical features of a brachiopod: side view (top) and dorsal view (not same genus)

Atrypa (ventral view)

2 ins

Spirifer (dorsal view)

"Articulate" brachiopods
Spiriferids

Spiriferids are defined by their internal spiral structure (*spiralium*) and are very variable externally. Occasionally the spiralium may be visible on a broken or weathered specimen. Non-spiriferid brachiopods may be placed with their major group on the basis of a few simple external features.

Spirifer *Carboniferous: Worldwide*
Relatively wide and strongly biconvex; hinge line long. Wide, long interarea on ventral valve only. Beak of ventral valve strong. Strong sulcus on ventral valve and fold on dorsal valve. Ornamentation of strong ribs which fork and are present on the fold and sulcus. Growth lines may also be present. Foramen absent.

Eospirifer *Silurian–Devonian: Worldwide*
Biconvex but ventral valve not very deep. Beak strong and ventral valve interarea almost horizontal. Hinge line long but less than maximum width of shell. Strong fold on dorsal valve and sulcus on ventral valve. Ornament of fine radiating ribs and concentric growth lines.

Atrypa *Silurian–Devonian: Worldwide*
Medium-sized. Dorsal valve very convex, ventral valve flattened or shallowly convex flexing downwards at its edges. Interareas absent but hinge line long or short. Beak small and turned inward. Ornamentation of ridges crossed by equally strong growth lines. Strong fold on dorsal valve and sulcus on ventral valve, particularly in old individuals.

Spirifer (ventral view)

2 ins

Eospirifer (dorsal view)

2 ins

Athyris (dorsal view)

Cyrtia (dorsal view)

Platystrophia (ventral view)

Cyrtia: side view (top) and front edge

Athyris *Devonian–Triassic: Worldwide*

A small to medium-sized spiriferid with a biconvex shell. Interareas absent, hinge line short. Shape varying from wide to elongate. Fold on dorsal valve and sulcus on ventral valve, both single smooth curves of variable strength. Beak strong and foramen present. Ornament of growth lines which may have the form of thick lamellae.

Cyrtia *Silurian–Devonian: Worldwide*

Medium-sized. Ventral valve (a) convex and very deep. Fold on dorsal valve and sulcus on ventral valve. Interarea of ventral valve very large (b) and almost vertical with high triangular projection in center. Shell surface smooth or carrying fine ridges and grooves.

Orthids

Orthis (dorsal view)

Hinge line long and interareas present on both valves. Shells biconvex.

Orthis *Cambrian–Ordovician: Worldwide*

Small to medium-sized. Ventral valve convex, dorsal valve shallowly convex or flattened. Hinge line equalling greatest width of shell. Interarea of ventral valve large; interarea of dorsal valve narrow. Interareas curve inward and both have triangular swellings or depressions near the middle. Ornament of strong radiating ribs. Dorsal valve usually with weak sulcus.

2 ins

Schizophoria (dorsal view)

Dalmanella (ventral view)

Platystrophia *Ordovician–Silurian: Worldwide*
Large to medium-sized. Strongly biconvex. Hinge line may equal greatest width, produced as point or sharp corner at each end. Interareas large, almost equal in size. Beak curving inward. Strong fold on dorsal valve and sulcus on ventral valve. Ornament of radiating ribs. Externally *Platystrophia* is indistinguishable from the spiriferids and is distinguished only by its internal structure.

(ventral view)

Schizophoria *Upper Silurian–Permian: Worldwide*
Medium-sized. Dorsal valve more convex than ventral valve. Interarea of ventral valve larger than that of dorsal valve; interareas shorter than hinge line which is less than greatest width. Low fold on dorsal valve and sulcus in ventral valve. Ornament of fine ribs and growth lines.

(ventral view)

Dalmanella *Ordovician–Silurian: Worldwide*
Medium-sized. Almost circular in outline. Dorsal valve more convex than ventral valve. Interarea of ventral valve long with curved surface which slopes downward. Interarea of dorsal valve shorter and curving upward. Weak sulcus sometimes on dorsal valve. Ornamentation of fine ribs of variable thickness. Growth lines strong near edges of valves.

(dorsal view)

Dicoelosia *Ordovician–Devonian: Worldwide*
Small to medium-sized. Strong sulci on both valves produce deep indentation on front edge. Hinge line shorter than greatest width. Interarea of ventral valve longer than that of dorsal valve. Ornamentation of ribs and growth lines.

(dorsal view)

Dicoelosia

Strophomenids

Interareas present on both valves; one valve usually convex and other concave.

Strophomena *Ordovician: Worldwide*
Dorsal valve (a) convex, ventral valve (b) concave. Hinge line long, corresponding to greatest width of shell. Interarea of ventral valve wider than that of dorsal valve. Triangular swellings in middle of upper and lower interareas. Ornament of fine radiating ribs.

Chonetes *Devonian–?Lower Carboniferous: Worldwide*
Dorsal valve (a) concave, ventral valve (b) convex. Hinge line long but not always widest part of shell. Surface with fine radiating ribs. Interarea of dorsal valve smaller than that of ventral valve. A row of spines is present along the edge of the interarea on the ventral valve; this feature is characteristic of the group to which *Chonetes* belongs.

Rafinesquina *Ordovician: Worldwide*
Ventral valve shown here. Large to medium-sized. This form is like *Strophomena* but with reversed convexity, that is the dorsal valve (a) is concave and the ventral valve (b) is convex. Hinge line long and small foramen on beak of ventral valve. Ornament of radiating ribs of variable thickness with the stronger ribs reaching to the beak. Middle rib of ventral valve usually very strong (shown here).

Strophomena: side view showing pedicle and brachial valve curvature

Chonetes: side view showing pedicle and brachial valve curvature

Rafinesquina: side view showing pedicle and brachial valve curvature

Strophomena (dorsal view)

Rafinesquina (ventral view)

Chonetes (dorsal view)

2 ins

Sowerbyella *Ordovician–Silurian: Worldwide*
Small to medium-sized. Dorsal valve concave, ventral convex. Hinge line corresponds to greatest width of shell. Ornamentation of fine radiating ribs.

Leptaena *Ordovician–Devonian: Worldwide*
Dorsal valve concave, ventral valve convex. Hinge line equals greatest width of shell and carries long, narrow interareas. Shells have very strong concentric ridges (rugae) and finer radiating ribs.

Productella *Upper Devonian–Lower Carboniferous: Eurasia*
Small to medium-sized, hemispherical to almost square shell with deeply concave dorsal valve (not shown here), and very convex ventral valve. Interareas very narrow, straight and poorly developed. Small spines scattered over ventral valve, but not on dorsal valve.

Spinulicosta *Devonian: Worldwide*
Small to medium-sized and similar to *Productella* to which it is very closely related. The shell is more elongate in *Spinulicosta* and carries an ornament of weak radiating ribs. Long slender spines may be present but are often not preserved. Interareas very narrow and straight as in *Productella*. Dorsal valve (not shown here) is dimpled and may carry concentric grooves.

Sowerbyella (dorsal view)

Leptaena

Productella (ventral view)

Spinulicosta (ventral view)

2 ins

2 ins

Productus

Sieberella
(dorsal view)

Conchidium (lateral view)

Productus
Lower Carboniferous: Eurasia N.Af China ?NA
Large. Ventral valve (shown here) highly convex and overlapping the hinge line. Dorsal valve flat. Ornament of radiating ribs. Spines may be scattered over the surface and two rows of spines on the ventral valve near the hinge line.

Pentamerids

Interareas present on both valves; shells biconvex; hinge line short.

Sieberella *Silurian–Devonian: NA E Af Asia*
Medium-sized and similar in general form to *Conchidium*, but with ventral valve usually even more convex. Beak very strong. Sulcus on dorsal valve and fold on ventral valve strong and carrying an ornamentation of ribs, but the rest of the shell surface is smooth. Commisure with a single, strong, angular curve.

Conchidium *Silurian–Devonian: Worldwide*
Large. Both valves very convex, ventral valve more so than dorsal valve. Beak of ventral valve curves upward and overlaps the beak of the dorsal valve (shown here in side view). Interarea of ventral valve small and interarea of dorsal valve obscured by inwardly flexed beak. Ornament of strong ribs. Fold and sulcus not developed. Commisure straight or with shallow curve.

Terebratulids

Interareas on ventral valves only, if visible. Shell surface usually smooth and foramen clearly visible on beak.

Dielasma *Carboniferous–Permian: Worldwide*
Small to medium-sized. Biconvex, shell surface smooth. Shell elongate, tear-drop shaped. Commisure may show a single curve which may be only feebly developed as shown. Foramen open and beak pointing upward and outward.

Dielasma: anterior view

Gibbithyris *Cretaceous: E Asia*
Medium-sized, biconvex. Commisure showing double curve as shown. Foramen open and beak pointing upward, or upward and inward. Shell surface smooth and outline less elongate than that of *Ornithella*.

Gibbithyris: anterior view

Ornithella *Jurassic: E Af*
Small to medium-sized, bi-convex with a smooth surface and weak or strong growth lines. Outline an elongate oval and commisure having an upward curve which is depressed centrally. Foramen clearly visible and beak pointing upward and outward.

Sellithyris *Cretaceous: E Af*
Medium-sized. Body flattened and biconvex. Shell surface smooth with strong growth lines. Commisure complex as shown and similar to that of *Gibbithyris*. Foramen open and large. Beak pointing upward, or upward and inward.

Sellithyris: anterior view

2 ins

Dielasma
(dorsal view)

Ornithella
(dorsal view)

Gibbithyris (dorsal view)

Sellithyris (dorsal view)

Rhynchonellids

Interareas very small or not visible. Shell surface with strong ribs, usually angular. Beak usually strong.

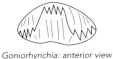

Goniorhynchia: anterior view

Goniorhynchia *Jurassic: E*
Medium-sized, biconvex, wider than long. Commisure as shown with single, strong, angular upward curve. Sulcus of dorsal valve and fold of ventral valve strongly developed. Beak strong and pointing upward and outward. Ornamentation of strong sharp-edged ribs giving the line of contact between valves a strong zig-zag appearance.

Cyclothyris *Cretaceous: NA E Af*
Relatively large rhynchonellid, similar in general form to *Goniorhynchia* but wider and more flattened. Upward fold of commisure weaker than in *Goniorhynchia*. Beak pointing upward.

Rhynchotrema *Ordovician: NA*
Small, biconvex. Sulcus of dorsal valve and fold of ventral valve well developed. Zig-zag commisure. Ornamentation of very strong ribs. Beak strong.

Hypothyridina *Devonian: Worldwide*
Large to medium-sized. Shell very high and biconvex. The commisure is characteristic, as the ventral valve is produced upward as a strong process which meets the dorsal valve near the top surface of the shell. The sulcus on the ventral valve and fold on the dorsal valve are well developed. Ornamentation smooth near beak but strong ribs near the front.

2 ins

Rhynchotrema (dorsal view)

Hypothyridina (dorsal view)

Goniorhynchia
(dorsal view)

Cyclothyris (dorsal view)

Hypothyridina (anterior view)

"Inarticulate" brachiopods

Valves not firmly joined. Interareas and hinge teeth never present.

Lingula *?Ordovician–Recent: Worldwide*
Elongate and nearly oval with small pointed hinge region. Shallowly biconvex and valves usually found separated. Ornament of numerous fine growth lines. Shell very thin with slight thickening near hinge and may have appearance of mother-of-pearl. (Many species have been placed in this genus. Together, their range extends from the Lower Paleozoic to Recent, but this long range for a single genus is unlikely and these species probably represent more than one genus. True *Lingula*, as typified by *L.anatina*, probably arose in the Cenozoic.)

Petrocrania *Ordovician–Devonian: E NA Asia*
Small. Usually found attached to other fossils by all parts of ventral valve, which is completely cemented to attachment surface. Shell conical or flattened and may carry radiating ribs as well as concentric growth lines. Four specimens of *Petrocrania* are shown here, arrowed, on the surface of an Ordovician strophomenid.

Petrocrania (dorsal view)
attached to a strophomenid

Lingula

2 ins

Bryozoans

2 ins

Colonial animals, mostly marine and important as fossils in many limestone deposits of Ordovician and later age. These typically delicate fossils may also be collected from weathered surfaces or washed from clays. Treatment of limestones with a weak solution (3 per cent) of hydrochloric acid allows good quality specimens to be recovered where silicification has occurred. Each individual *zooid* of the colony builds a calcareous tube or box known as a *zooecium*, and the skeleton of the colony as a whole is called the *zoarium*. The opening of each zooecium is an *aperture*. Knowledge of microscopic morphology is needed for precise identification in most cases. This usually entails study of thin sections for Paleozoic species, and well-preserved colony surfaces for Mesozoic and Cenozoic species. Thin sections may reveal the presence of transverse partitions (*diaphragms*) in the zooecia. Growth form varies greatly between species. The naming of growth forms used here is the same as that given for corals (*page 226*). Certain massive and thick branched forms are very coral-like, and some groups were once treated as corals. Six subgroups (orders) of bryozoans are mentioned here: Cryptostomata, Fenestrata, Trepostomata, Cystoporata, Cyclostomata, and Cheilostomata. The first four are mostly or entirely of Paleozoic age, whereas the Cyclostomata range from the Ordovician to Recent, and the Cheilostomata from the Jurassic to Recent.

Archimedes

Cryptostomata

Apertures rectangular or polygonal, regularly arranged on colony surface. Zooecial tubes short to moderately long. Colonies erect, either flattened bifoliate fronds or narrow cylindrical branches, occasionally jointed.

Ptilodictya *Ordovician–Devonian: NA E Asia*
Fronds sickle-shaped, tapering basally. Cross-section of fronds oval or diamond-shaped with a median wall. Apertures rectangular and arranged in lines.

Fenestrata

Apertures opening only on one side of the narrow colony branches. Zooecial tubes short. Colonies erect, often net- or fern-like.

Fenestella *Ordovician–Permian: Worldwide*
Net-like zoarium forming a planar fan or a funnel. Branches bear two rows of zooecial apertures separated by a central keel and are linked by narrower cross-bars lacking apertures.

Fenestella: two rows of apertures

Archimedes *Carboniferous–Permian: NA Asia*
Easily identified from the spiral, screw-like axis of the colony (shown here), usually the only part to be preserved. In complete colonies the axis carries a twisted net-like frond which is virtually indistinguishable from *Fenestella* when dissociated.

Polypora: more than two rows of pores

Polypora *Ordovician–Permian: Worldwide*
Like *Fenestella* but having the apertures arranged in more than two rows without a central keel.

Penniretepora *Devonian–Permian: Worldwide*
Delicate, fern-like colony with primary branches bearing regularly-spaced, short side branches. Apertures arranged in two rows separated by a central keel.

Penniretepora: surface features

2 ins

Fenestella

Polypora

Ptilodictya

Penniretepora

2 ins

Trepostomata

Apertures polygonal, sometimes with two size classes, not regularly arranged on the colony surface. Zooecia tubular, long, typically thin-walled initially in colony interior but becoming thick-walled toward exterior. Colony massive, branching, frond-like or encrusting. Surface often covered with small swellings known as *monticules.*

Monticulipora *Ordovician: NA*
Growth form massive (shown here); less commonly branching or frond-like. Monticules well-developed. Zooecia having large apertures surrounded by others with small apertures. This and other "monticuliporoids" were once thought to be corals.

Cystoporata

Apertures circular, usually with a hooded structure (*lunarium*) projecting over one side. Zooecia tubular, separated by areas of cyst-like calcification (*cystopores*) in most species. Colony encrusting, frond-like, massive or branching.

Fistulipora *Silurian–Permian: Worldwide*
Zoarium usually encrusting but may be massive or branching, sometimes forming large sheets up to 12 inches across (a piece of such a sheet is shown). Large zooecial apertures subcircular in shape and separated by cystopores.

Constellaria *Ordovician: NA E*
Growth form of cylindrical or flattened branches. Colony surface covered by distinctive star-shaped monticules.

Fistulipora

Monticulipora

2 ins

Cyclostomata

Apertures circular or polygonal, rarely semicircular. Zooecia tubular, invariably with porous walls. Large polymorphic zooecia for larval brooding present in most species.

Meliceritites *Cretaceous: E*
Zoarium consisting of slender bifurcating branches. Zooecia at colony surface are hexagonal with a semicircular aperture closed by a hinged cap (operculum) which may be lost.

Meandropora *Pliocene: E*
Massive zoarium consisting of radiating cylindrical bunches (fascicles) of tubular zooecia, united at intervals or linked by shelves (shown here). Polygonal apertures visible at the ends of the fascicles.

Blumenbachium *Pliocene: E*
Similar in overall colony shape to *Meandropora*, but with more complex colonies consisting of multiple layers of coalesced subcolonies which are polygonal on the colony surface and often form raised ridges at their junctions. Apertures are small and polygonal.

2 ins

Constellaria

Meliceritites

Meandropora

Blumenbachium

Stomatopora *Triassic–Recent: Worldwide*
Encrusting, thread-like, zoaria consisting of narrow, bifurcating branches, one zooecium in width. Apertures are circular and spaced regularly along the branches.

"Berenicea" *Triassic–Recent: Worldwide*
Zoarium small, usually less than 0.5 of an inch in diameter and typically circular, forming a thin, encrusting sheet. Zooecia with circular or oval apertures distributed evenly over the colony surface. The specimen shown here encrusts a plate of a sea urchin.

Reticrisina *Cretaceous: E*
Erect zoarium consisting of a network of compressed branches. Apertures circular and arranged in raised rows on the sides of the branches.

Cheilostomata

This group includes the commonest living bryozoans. Growth forms are variable, and include delicate branching, net-like, frondose and sheet-like encrusting zoaria. Zooecia are typically box-shaped. Apertures are usually not circular, and in living forms each aperture is closed by an operculum which is generally uncalcified. Specialized zooecia (*avicularia*) have modified opercula enlarged into mandibles used defensively.

Callopora *Cretaceous–Recent: Worldwide*
Encrusting zoaria of irregular shape with zooecia arranged in regular rows. Aperture oval and occupying most of the surface of the zooecium, ringed by small circular holes representing the bases of articulated spines. Avicularia present, visible as small diamond-shaped openings with crossbars in well-preserved examples. Specimen shown here is attached to a sea urchin plate.

Onychocella *Cretaceous–Recent: Worldwide*
Encrusting or erect, forming sheets of variable size. Apertures approximately semicircular, located at the end of a depressed frontal wall. Zooecia commonly hexagonal in shape, often appearing slightly to overlap one another. Specimen shown here is attached to a bivalve mollusk shell.

Lunulites *Cretaceous–Recent: Worldwide*
Small zoarium, usually less than 0.5 of an inch diameter, in the shape of a low, flattened cone. Apertures of a similar shape to *Onychocella* and arranged in regular radial rows separated by grooves containing avicularia. Concave underside lacks apertures.

2 ins

"Berenicea"

Reticrisina

Stomatopora

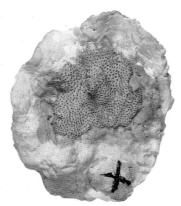

Onychocella

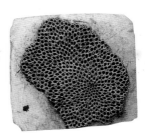

Callopora

Lunulites

Echinoderms

Fossil echinoderms are easily identified because of their characteristic five-fold symmetry and their skeleton of many plates, each
of which is composed of a single calcite crystal. Six groups, sea
lilies, blastoids, starfishes, brittle stars, edrioasteroids and echi-
noids are mentioned here. The sea lilies and sea urchins are the
most important as fossils.

Echinoderm groups

Plant-like with small cup, conical net of feather arms and usually also with stem of diskoidal plates.	
A Arms arise as outgrowths from cup	**sea lilies**
B Bud-like cup with five radiating grooves	**blastoids**
Diskoidal, with five raised, often sinuous grooves	**edrioasteroids**
Star-shaped body form	
A Arms thin and separated from central disk	**brittle stars**
B Arms merging gradually into body	**starfishes**
Globular, hemispherical or heart-shaped shell with two openings and covered in tubercles and spines.	**sea urchins**

Sea lilies (Crinoidea)

Plant-like body consisting of a *cup* from which arise five branching arms which form a filtration cone. Fine side branches from the
arms are known as *pinnules*. The cup is usually supported by
a stem composed of diskoidal elements stacked on top of
each other. Often stem ossicles are found in isolation (*shown in
Cyathocrinites on page 298.*) Some species have become secon-
darily free-living and lack a stem (*as in Marsupites on page 298,
and Uintacrinus, below*).

Sagenocrinites Silurian: NA E
Cup large, composed of many hexagonal plates and incorporating the lower parts of the arms. Arms branching dichotomously,
without pinnules. Stem circular in cross-section.

Taxocrinus Devonian–Carboniferous: NA E
Cup relatively short and including lower parts of arms; but arm
ossicles clearly differentiated from cup plates, the latter being
much smaller than in *Sagenocrinites*. Arms branching dichotomously, without pinnules. Stem circular in cross-section.

Uintacrinus Upper Cretaceous: NA E Aust
Cup large, composed of many small hexagonal plates, with lower
parts of arms bound into cup. Arms branch once and bear pinnules. No stem.

Pentacrinites *Jurassic: NA E Asia*

Cup tiny, with long, highly branched arms bearing pinnules. Arms entirely free of cup. Stem very long, composed of star-shaped ossicles often found in isolation. The stem has fine lateral tendril-like processes termed *cirri*.

2 ins

Taxocrinus

Pentacrinites

Sagenocrinites

Uintacrinus

Marsupites *Cretaceous: NA E Asia, Aust, Af*
Cup large, composed of a small number of large polygonal plates arranged in three circlets. Arms almost completely free of cup; dichotomously branching and with pinnules. No stem.

Cyathocrinites *Silurian–?Permian: Worldwide*
Cup small, bowl-shaped, composed of three circlets of large plates. Arms well-separated and free of cup, branching dichotomously; lacking pinnules. Stem circular in cross-section. Several discoidal stem elements can be seen scattered among the arms and to the top left of the specimen shown here. These have a wide central hole and a marginal arrangement of radiating ridges and grooves.

Phanocrinus *Carboniferous: NA E Af*
Cup small and bowl-like, composed of two circlets of five polygonal plates plus three small plates in basal concavity. Arms free of cup, relatively short and branching once immediately above the cup; bearing pinnules. Stem long and circular in cross-section.

2 ins

Marsupites

Phanocrinus   *Carpocrinus*   *Cyathocrinites*

298

2 ins

Platycrinites

Glyptocrinus

Dichocrinus

Platycrinites   Devonian–Permian: NA E Asia

Cup large, composed of just two circlets of large polygonal plates, the upper with five plates, the lower with three. Arms free of theca and branching two or three times close to base; composed of a single series of plates near their base but becoming double (biserial) distally; bearing pinnules. Stem ovate in cross-section and characteristically helically twisted.

Glyptocrinus   Ordovician–Silurian: NA

Cup large, conical, composed of a large number of plates ornamented by strong radial ridges, and incorporating the lower parts of the arms. Arms branching dichotomously twice; bearing pinnules. Stem cylindrical with pentagonal central perforation.

Platycrinites: biserial arm

Carpocrinus   Silurian: NA E

Cup moderately large, conical and incorporating the lower parts of the arms; plates polygonal. Base of cup composed of three approximately equal-sized plates (a, b, and c). Arms branching once in cup so that there are ten strong free arms bearing pinnules. Stem circular in cross-section.

Carpocrinus: base of theca

Dichocrinus   Carboniferous: NA E

Cup bowl-like, composed of two circlets of large plates, the lower circlet composed of just two plates. Arms generally branched just once and composed of double elements except close to base where they are single. Pinnules on arms long and well-developed. Stem circular in cross-section. Here *Dichocrinus* lies on top of a specimen of *Rhodocrinus*, which is similar in general appearance but has a cup composed of numerous small plates.

Dichocrinus: base of theca

299

Starfishes (Asteroidea)

Star-shaped body composed of many small platelets loosely bound together, usually with five arms which are not sharply marked off from the body. Some have large skeletal plates edging the body, termed *marginal plates*. Plates running down the mid-line of each arm on the oral surface are termed *ambulacral plates*. Complete starfishes are rare as fossils, but where conditions are right for their preservation they are commonly abundant.

Pentasteria *Jurassic–Eocene: E*
Arms rather straight and subparallel-sided with strong marginal plates forming double series. Central body area rather small; covered in small granular platelets on upper surface. Ambulacral plates slender with large gaps.

Mesopalaeaster *Ordovician: NA E*
Arms narrow, bounded by a single series of marginal plates. Aboral surface (not illustrated) with distinct rows of stellate plates aligned along the arms. Ambulacral plates block-like without gaps.

Calliderma *Cretaceous–Recent: Worldwide*
Cushion-shaped with short projecting arms and strong frame composed of a double series of marginal plates. Upper surface covered in tessellated platelets with semi-regular arrangement.

Brittle stars (Ophiuroidea)

Star-shaped body with narrow, cylindrical arms clearly separated from a circular disk-like body. Ophiuroids are distinguished on details of their disk and arm plating and are not readily identifiable without a microscope.

Palaeocoma *Jurassic: E*
A typical brittle-star with long, flexible, whip-like arms and a small disk-shaped body.

Edrioasteroids (Edrioasteroidea)

Diskoidal to subglobular echinoderms that attach directly to hard substrata. Most have only the upper surface plated which comprises an outer marginal ring and a central plated body with five sinuous grooves (*ambulacra*) radiating from a central mouth.

Edrioaster *Ordovician: NA*
Hemispherical with ambulacra extending to outer edge (*ambitus*) and curved around body. Remaining plates polygonal and tessellate. Marginal frame hardly differentiated and on underside.

Blastoids (Blastoidea)

Attached, plant-like fossils with stem, bud-like body, and fine feathery appendages termed *brachioles*. The body has five prominent grooves radiating from an apically positioned opening (the mouth) and is composed of a small number of large polygonal plates, including five V-shaped plates associated with each groove. Usually only the plated body is preserved.

Pentremites

Pentremites *Carboniferous: NA*
The specimen illustrated shows just the bud-like body. *Pentremites* is one of the commonest blastoids.

2 ins

Mesopalaeaster

Pentasteria

Edrioaster

Calliderma

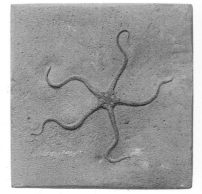

Palaeocoma

301

Sea urchins (Echinoidea)

Globular, ovate, hemispherical or heart-shaped animals with a *test* (body) formed of vertically aligned rows of polygonal plates sutured together. These plates are covered in *spines*, which are only loosely attached and commonly become detached before fossilization. *Tubercles* over the plate surface mark the sites of attachment of these spines. *Ambulacral plates* are perforated by *pores* which form easily recognizable radiating tracts. In between ambulacra are two columns of *interambulacral plates*. The test has two openings, the *mouth* and *anus*.

Regular echinoids

Mouth and anal openings are situated at opposite poles of the test. Tubercles are generally prominent covering all plates. The structure of tubercles, whether with a central perforation and whether bearing cog-like crenulation, is important for identification.

Pedina Jurassic–Miocene: NA SA E Madagascar
Subglobular test with rather sparse covering of small tubercles that are perforate and lack crenulation. Anal opening central within a circle of five large plates.

Pedina

2 ins

Psammechinus

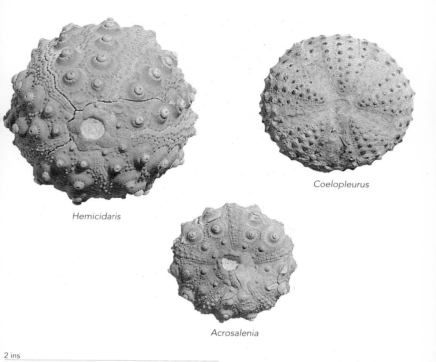

Coelopleurus

Hemicidaris

Acrosalenia

2 ins

Psammechinus *Pliocene–Recent: NA E Af*
Depressed test covered in rather uniform tubercles which are imperforate and lack crenulation. Plates around anal opening usually lost. Pores offset in arcs of three along the ambulacra.

Acrosalenia *Jurassic–Cretaceous: Worldwide*
Test with single large tubercle to each interambulacral plate which is perforate and crenulate. Mouth opening large; anal opening small and displaced within apical cycle of plates imparting a strongly bilateral symmetry. Ambulacra narrow, often sinuous.

Hemicidaris *Jurassic: NA E Af Asia*
Like *Acrosalenia* but with anal opening central within apical circlet of plates. Tubercles on ambulacral plates decrease markedly in size on upper surface.

Coelopleurus *Eocene–Recent: Worldwide*
Test depressed with five wide aboral zones free of tubercles. Tubercles imperforate and without crenulation. Pores in single series above but crowded and forming broad zones toward large mouth opening.

Irregular echinoids

Anal opening displaced from the apex of the test; usually posterior on the oral surface. Tubercles generally very fine and uniform. Most have aboral pores double and strongly elongated to form petals.

Pygaster *Jurassic: E*
Test depressed with large central mouth on underside notched around its edge. Anal opening very large and key-hole-shaped on upper surface, but displaced to the posterior of the apex. Tubercles clearly seen and slightly sunken. Pores circular and not petaloid.

Micraster *Cretaceous–Paleocene: Worldwide*
Heart-shaped with mouth on lower surface near the base of the frontal groove and anal opening on the posterior surface. Petals differentiated and weakly sunken on upper surface. Plates at the apex compact. Two large plates covered in large tubercles form much of the test behind the mouth.

Holaster *Cretaceous: Worldwide*
Heart-shaped, like *Micraster*, but without sunken petals and with apical plates stretched out along the anterior-posterior axis.

Clypeaster *Eocene–Recent: Worldwide*
Test very thick-shelled; pentagonal to oval in outline and domed or subconical in profile. Mouth small and central on lower surface; somewhat sunken with five grooves radiating from it. Petals strongly developed on upper surface. Anal opening small, at posterior margin of lower surface.

Conulus *Cretaceous: Worldwide*
Conical in profile. Mouth small, circular and central on flat lower surface. Anal opening just below margin at rear of lower surface. Ambulacral pores simple throughout with no petal development. Commonly found as internal casts in flint (shown here) derived from the Chalk of Europe.

Pygurus *Jurassic–Cretaceous: Worldwide*
Test depressed and pentagonal in outline. Small pentagonal mouth opens a little anterior of center on the lower surface; the anal opening is also on oral surface, at the posterior border. Petals strongly developed aborally.

Echinolampas *Eocene–Recent: Worldwide*
Test ovate with small pentagonal mouth slightly anterior of center, and anal opening wider than long and visible in oral view at the posterior border. Petals well-developed with unequal-lengthed rows of pores in each ambulacrum.

Echinolampas

Pygaster

Holaster

Clypeaster

Conulus

Pygurus

Micraster (ventral view)

2 ins

Micraster (dorsal view)

Graptolites

A group of colonial, usually planktonic animals. The class is an extinct group of hemichordates, with distant vertebrate affinites. Its members were important and common from the Cambrian to the Carboniferous. Graptolites are important for dating Paleozoic rocks as they changed very rapidly through time and many genera had worldwide distribution. Graptolites are common in shales and slates in which they are flattened along the bedding planes and are usually carbonized. They may be difficult to see on the rock surface, but by slanting the specimen to the light they are usually seen to have shiny surfaces.

Each graptolite colony is known as a *rhabdosome* and consists of a variable number of branches or *stipes* that diverge from the initial individual of the colony, which is known as the *sicula*. The *nema* is the thread-like process by which the rhabdosome may be attached. Each individual of the colony is housed in a cup-like structure known as the *theca*.

Diplograptus: biserial growth form

Monograptus: Monoserial growth form

Tetragraptus: pendent form

Dicellograptus: growth form

Dendrograptus   *Cambrian–Carboniferous: Worldwide*
An attached, plant-like form. Rhabdosome consisting of numerous stipes which give it a fern-like appearance.

Diplograptus   *Lower Ordovician–Lower Silurian: Worldwide*
Member of the graptoloid group (Graptoloidea) of the graptolites, which also includes *Monograptus*, *Dicellograptus*, and *Tetragraptus* (see below). Members of this group were important planktonic forms in the Ordovician and Silurian. *Diplograptus* has thecae arranged on each side of the stipes as shown (that is, it is *biserial*). This shows as double serrations on the ribbon-like specimens.

Monograptus   *Lower Silurian–Lower Devonian: Worldwide*
Thecae arranged in a single row along the side of the stipe as shown (that is, it is *monoserial*). Rhabdosomes may be coiled, spiral or straight, and fragmentary stipes of the other genera in the family (Monograptidae) also resemble *Monograptus*.

Tetragraptus   *Lower Ordovician: Worldwide*
Rhabdosome consisting of two short branches that diverge from the nema and then fork again quickly into two more, making four, as shown. Each branch has thecae on one side only. Serrated edges are clearly shown on this specimen.

Dicellograptus   *Lower–Upper Ordovician: Worldwide*
Consisting of two stipes that are characteristically flexed from the center as shown, and carry thecae on one side only.

2 ins

Dendrograptus

Diplograptus

Monograptus

Tetragraptus (reclined form)

Dicellograptus

307

Vertebrates

Fishes, amphibians, reptiles, birds, and mammals. Vertebrates have internal skeletons of cartilage or bone. Complete fossil skeletons are rare; isolated bones or teeth being much more common.

Fishes

The largest group of living vertebrates with over 20,000 species and a huge number of fossil forms.

Armored fishes

Many Paleozoic fishes had a heavy external armor of bone. These are usually found isolated. Fossils are usually of Silurian and Devonian age.

Cephalaspis *Silurian–Devonian: NA E Asia*
One of the best known armored fishes. A complete specimen is shown. Note the wide head which is covered with bony plates and slender tail covered with scales.

Coccosteus *Middle and Upper Devonian: Eu Asia NA*
A skull roof is shown; shield-shaped, dorso-ventrally compressed, and composed of paired and median bony plates, joined at sutures and ornamented with tubercles. The lateral plates at the back of the skull roof are modified to articulate with the bony trunk shield.

2 ins

Cephalaspis

Coccosteus
(skull roof in a concretion)

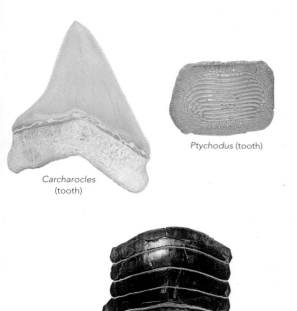

Carcharocles
(tooth)

Ptychodus (tooth)

Myliobatis (tooth plate)

2 ins

Hybodus (spine)

Sharks and rays

Skeleton composed of cartilage and rarely fossilized. Teeth are quite common in the Carboniferous, becoming more common in the Cretaceous and Tertiary when cartilaginous fish diversified.

Hybodus *Triassic–Cretaceous: Worldwide*
Teeth low and wide, high central point and numerous side points. Spine long and pointed with grooved sides. These spines support the dorsal fins in members of the hybodont group of sharks which were common during the Mesozoic.

Carcharocles *Paleocene–Pleistocene: Worldwide*
Very large teeth with a single point and serrated edges.

Ptychodus *Cretaceous: NA E Af Asia*
Flattened teeth suitable for crushing mollusk shells. This is a hybodont shark, but its teeth are similar to those of many rays.

Myliobatis *Cretaceous–Recent: Worldwide*
A ray. Flattened crushing teeth indicating a shellfish diet.

Bony fishes

Include most living fishes such as the salmon, cod, and herring. The group was important in fresh water by the end of the Paleozoic, and has since become important in marine conditions. Identification is very difficult.

Cheiracanthus

Middle Devonian: Europe; Lower or Middle Devonian: NA
An articulated fish; body covered with tiny scales, smooth, or ornamented with faint longitudinal ridges. A long, slender fin spine supports the anterior margin of each fin. The shoulder girdle is preserved here, just above the base of the pectoral fin spine, just behind the head.

2 ins

Cheiracanthus
(in a concretion)

Osteolepis

Brookvalia (on a bedding plane)

Ceratodus (tooth plate)

Perleidus (in a nodule)

2 ins

Osteolepis *Middle Devonian: Europe*
An articulated, lobe-finned fish. The scales are rhombic. The two dorsal fins and the anal fin are each separate. The paired fins had a short, fleshy lobe at the base. All fins have an internal skeleton at the base and slender, bony, straight dermal rays (lepidotrichia). The tail is heterocercal.

Ceratodus *Triassic–Paleocene: Worldwide*
Lung fish, generally known from fossil teeth only. Shape and ridges are characteristic. Surface with many small pores.

Perleidus *Triassic: E Af Asia*
Complete bony fishes may be found in nodules and the presence of the dead fish sometimes appears to have caused the formation of the nodule.

Brookvalia *Triassic: Aust*
Fossil fishes may be found flattened out along bedding planes and are discovered when the rock is split.

Reptiles

Include turtles, ichthyosaurs, plesiosaurs, lizards and snakes, crocodiles, pterosaurs, and dinosaurs. Reptiles were the dominant land animals from the Permian to the end of the Cretaceous and the top predators in Jurassic and Cretaceous seas.

Crocodiles

Triassic–Recent: Worldwide

Crocodiles are among the commonest fossil reptiles, but they are very difficult to identify generically. A bony scute and two teeth are shown here. The scutes are arranged in rows along the back of the animal, beneath the skin, and always have heavily pitted upper surfaces. Crocodile teeth vary greatly along the jaw of the same individual. They usually have short, sharply pointed crowns (the black upper part), and long roots.

Turtles

Jurassic–Recent: NA E Af Asia

Pieces of turtle shell or carapace are the commonest parts found. A plate from the upper part of the shell of *Trionyx*, a freshwater softshell turtle, is shown here. In freshwater turtles the plates sometimes have patterns on their upper surfaces but in marine turtles the surfaces of the plates are smooth. In life the plates had a horny covering.

Ichthyosaurs

Jurassic–Cretaceous: NA SA E Asia Aust

Rarer than crocodiles or turtles but important marine reptiles in the Mesozoic, especially the Jurassic. The most frequently found parts are the centra of the vertebrae. The two swellings at the top indicate where the neural arch was broken off. A skull fragment consisting of part of the upper and lower jaws and teeth is also shown here. The crowns of the teeth carry deep vertical grooves and the roots are bulbous and fluted.

Dinosaurs

Triassic–Cretaceous: Worldwide

Dinosaurs have been found on every continent. Their remains are most abundant in North America, China and Mongolia but at least 30 different dinosaurs are known from Britain.

Albertosaurus Cretaceous: NA

Flesh-eating dinosaurs have tall sharp, blade-like teeth with serrated edges. A single tooth of *Albertosaurus* is shown here.

Iguanodon Cretaceous: NA E Asia

Many plant-eating dinosaurs have square crowned teeth with flat upper surfaces and ridged sides. A single tooth of *Iguanodon* is shown here.

Hypsilophodon Cretaceous: E

Not all dinosaurs were large and a femur of *Hypsilophodon* is shown here. This plant-eating dinosaur was about 3 feet tall and 7 to 10 feet long.

2 ins

Trionyx (carapace plate)

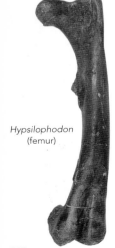

Hypsilophodon (femur)

Birds

The oldest known bird occurs in the Jurassic. Many different kinds are known from the Cretaceous although most modern-type birds appear in the Eocene. Their bones are very fragile as they have thin walls and an empty internal cavity; as a result they are only rarely preserved as fossils. You are most likely to find them in Pleistocene deposits and they can usually be identified by comparison with living bird bones. Shown here is the metatarsus (long part of the foot) of a Dodo (Pleistocene: Mauritius) which has a form characteristic of birds, as there are three articulating surfaces for the toes at the lower end; a feature not found in mammals or reptiles. Other bird bones may sometimes be confused with bones of mammals or reptiles.

Crocodile (teeth)

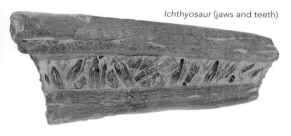

Ichthyosaur (jaws and teeth)

2 ins

Ichthyosaur (vertebra)

Dodo (metatarsus)

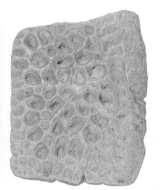

Crocodile (scute)

Albertosaurus
(tooth)

Iguanodon
(tooth)

313

Mammals

Includes the animals which are typically covered with hair and which suckle their young (like humans, horses, elephants, whales, bats, and dogs). The left and right side of the lower jaw are each composed of a single bone, the *dentary*. The teeth are socketed into the jaw and usually well differentiated into distinct functional types (*incisors, canines, premolars, molars*). A major group of animals since the end of the Cretaceous, remains are common in Pleistocene deposits and locally abundant in some earlier deposits, for example the Oligocene of South Dakota. The teeth are often good indicators of the type of food that was eaten and are very important in the identification of most mammals.

Equus (upper molar)

Flesh eaters

Teeth usually modified into either piercing points or shearing blades aligned along the axis of the jaw and are elongate in plan view. The blades often have marked V-shaped notches.

Canis (wolf, domestic dog, jackal, dingo)
Miocene–Recent: Worldwide
One of the cheek teeth in each jaw is large and elongate with a sharp notched edge that is used for slicing flesh. Cats, hyaenas, weasels, and civets also have generally similar slicing teeth.

Equus (lower molar)

Adcrocuta (extinct hyaena) *Miocene: E Af Asia*
This is a skull of a young individual, but it shows the long slicing cheek tooth and the relatively small number of teeth. The upper canines are not erupted, but the points are visible near the front of the jaw. The arch of the jaw is wide to accommodate large jaw muscles, and the face is relatively short. These are features of most flesh-eating mammals, but are highly developed in the hyaenas which are adapted for crushing bones.

2 ins

Canis (lower jaw fragment)

Adcrocuta (skull)

Plant-eating mammals – grazers

Eaters of rougher plant food including grasses. Teeth usually
have high crowns, are square to rectangular in plan view and have
relatively flat, rough, biting surfaces.

Bos (Cattle, including the domestic cow)
Pleistocene–Recent: Alaska E Af Asia
(Transitional between browser and grazer). Upper teeth with four
crescentic cusps forming square crown. Lower molars rectangular
with an extra cusp at the back of the last molar. Bison (NA),
antelopes and gazelles (E Af Asia), deer and giraffes (E Af Asia)
have cheek teeth with similar patterns.

Bos (crown view of upper molar)

Mammuthus *Pliocene–Pleistocene: E Af Asia; Recent: Asia*
(A fossil relative of the living Asian and African elephants.)
Very large cheek teeth consisting of wide, almost parallel-sided
platelets forming ridges on biting surface.

Bos (lower molar)

Equus (horse, donkey, zebra)
Pliocene–Recent: NA SA E Af Asia
Teeth very high with square crowns (upper in plate) and rectangular crowns (lower in plate). Pattern complex. In Eocene, for
example, *Pliolophus* (shown on page 317), Oligocene and early
Miocene horses have low crowned teeth similar to those of small
rhinoceroses.

Equus: crown view of upper molar tooth

Castor (beaver) *Pliocene–Recent NA E Asia*
Complete lower jaw shown. Front tooth extremely long with
almost triangular cross-section and enamel on front face only.
Cheek teeth few in number, separated from front tooth by a
space, very high crowned, flat-topped with several cross-crests.

2 ins

Mammuthus (molar)

Castor (lower jaw)

Plant-eating mammals – browsers (leaf-eaters)

Eaters of softer plant food and mixed feeders. The teeth usually have quite low crowns and are square to rectangular in plan view. The biting surfaces have well-developed cusps or crests for crushing and shearing.

Ursus (canine)

Ursus (upper molar)

Ursus *Pliocene–Recent: NA E Asia*
(Transitional between flesh-eater and plant-eater). Includes grizzly bear and brown bear. Cheek teeth with low crowns, low rounded cusps and many additional small bumps and grooves. Some pigs have similar cheek teeth. Canines large with swollen root and pointed crown. Diet also includes flesh.

Merycoidodon *Oligocene: NA*
Also known as *Oreodon* and very common in Oligocene of Mid-West, USA, where beds are known as 'Oreodon beds'. Skull relatively short and deep. A leaf-eating mammal having upper molars similar in general crown pattern to *Bos*; consisting of four crescents but crowns much lower. Upper canine relatively large. A sheep-sized relative of camels.

Rhinoceroses *Eocene–Recent: NA E Af Asia*
Upper teeth (upper in plate) with continuous outer walls and two inner crests. Lower teeth (lower in plate) consisting of two crescentic ridges.

Diprotodon *Pliocene–Pleistocene: Australia*
Upper teeth shown. These each have a pair of low sharp-edged cross-crests. *Diprotodon* is a marsupial and is therefore related to

2 ins

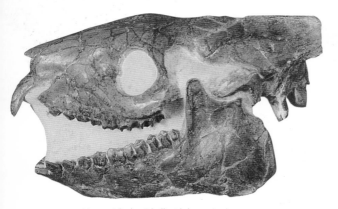

Merycoidodon (skull with lower jaw)

Rhinoceros (upper molar)

Rhinoceros (lower molar)

the kangaroo, koala, wombat and opossum. Remains of marsupials are the commonest mammalian fossils in Australia and also occur in SA (Paleocene onwards), NA (beginning in the Cretaceous) and E (Eocene–Miocene). There are rare occurrences in the Tertiary of Af Asia Antarctica. In the Miocene to Pleistocene of E Af and Asia, large teeth similar to *Diprotodon* are from the giant elephant-relative, *Deinotherium*, while smaller ones are from pigs or tapirs (not Af). *Pyrotherium* from the Oligocene of SA also had similar teeth.

Mammut *Miocene–Pleistocene: NA E Af Asia*
(Has nothing to do with the mammoth, in spite of its name.) Remains relatively common in North American Pleistocene; known as American mastodon. Cheek teeth large with several cross-crests but these are much lower and more triangular than in the elephantids. The enamel on this type of molar is very thick.

Hippopotamus *Pliocene–Recent: E Af Asia*
A lower molar shown. Four cusps arranged in a rectangle; similar in general pattern to *Bos* but cusps less crescentic. Some pigs have similar teeth, as do members of an extinct group, the anthracotheres.

Hippopotamus
(lower molar)

Pliolophus *Eocene: NA E*
Formerly known as *Eohippus* or *Hyracotherium*; the first horse. Skull long and low. Cheek teeth low crowned with four rounded cusps on the upper molars. You are unlikely to find remains of this animal but it is displayed in most museums.

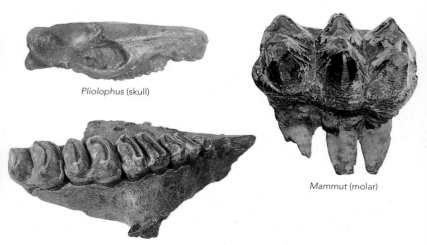

Pliolophus (skull)

Mammut (molar)

Diprotodon (upper teeth)

2 ins

Fossil Land plants

Land plants are common fossils, particularly in terrestrial sediments. Plants produce prodigious quantities of seeds, fruits, and pollen or spores, and many species will shed whole organs (e.g. leaves) either continuously or at various times of the year. Many of these plant parts are incorporated into the fossil record. Under certain conditions (e.g. coal swamps), plants are fossilized at their site of growth providing crucial information on the morphology of the whole plant and important insights into the ecology of ancient terrestrial ecosystems. The fossil record contributes information on climatic and ecological changes as well as data on plant evolution. Pollen and algae are widely used in dating certain types of rocks. A representative selection of fossil land plants is included here.

Zosterophylls

Devonian: NA SA E Asia Aust
Amongst the earliest known land plants, these fossils are most closely related to living clubmosses (lycopsids). Very simple plants lacking leaves, roots and seeds.

Sawdonia *Devonian: NA E Asia*
Simple branched stems with coiled tips in younger parts. Stems bearing conspicuous spines.

Coal Measure Plants

Coal is formed from plants, and the Coal Measures of the Carboniferous Period are a major source of fossil plants. Spoil heaps at coal mines are excellent places to collect. With the exception of flowering plants, many of the major living groups of land plants had evolved by the Carboniferous Period.

Lycopsids (clubmosses)

Late Silurian–Recent: Worldwide
Living lycopsids are relatively small herbaceous plants that are a minor component (<1%) of modern species diversity. The zenith of lycopsid evolution was the Carboniferous where as much as 50% of known fossils are attributable to the group. Unlike their living relatives, some extinct species were very large trees up to 100 feet in height.

Lepidodendron *Carboniferous–Permian: Worldwide*
Part of a branch from the apex of a large tree. Branch bears numerous, narrow, needle-like leaves. Lower branches and trunk are considerably wider and have a superficial pattern of diamond-shaped leaf bases.

Sphenopsids (horsetails)

Devonian–Recent: Worldwide
This group contains only 15 living species, and all are relatively small herbaceous plants. Sphenopsids have a lengthy and diverse fossil record, and many Palaeozoic species were large trees up to 70 feet in height. Plants in this group have characteristic jointed stems with branches and leaves in whorls.

Calamites *Carboniferous–Permian: NA SA E Asia Aust*
The stems of living and extinct sphenopsids have a hollow central region. This region may become filled with sediment during fossilization to produce an internal cast of the pith cavity with characteristic vertical ridges and joints.

Annularia *Carboniferous–Permian: E Asia*
Successive whorls of needle-shaped leaves from the terminal branches of the *Calamites* plant.

2 ins

Sawdonia (spiny stems)

Calamites
(internal stem cast)

Lepidodendron
(leafy branch)

Annularia

2 ins

Pecopteris   Ptychocarpus   Neuropteris

Fern-like foliage

Devonian–Recent: Worldwide
Pinnate leaves are a common element of Late Paleozoic and
Mesozoic floras. Foliage of this type is characteristic of living and
fossil ferns as well as extinct seed plants that are more closely
related to living cycads, conifers and flowering plants than to
true ferns. In the absence of reproductive structures, the precise
affinity of much fossil fern-like foliage is difficult to establish.

Pecopteris *Carboniferous–Permian: Worldwide*
Foliage typical of some extinct ferns (e.g. *Psaronius*) and seed
plants. Pinnules attached along entire width of base; with or without parallel margins; distinct vein extending almost to pinnule
tip.

Ptychocarpus *Carboniferous–Permian: Worldwide*
Foliage similar to *Pecopteris* but with spore-bearing organs consisting of numerous microscopic circular structures attached to
leaf surface (hand lens required) that demonstrate an affinity with
ferns rather than seed plants.

Neuropteris *Carboniferous: E NA SA Asia*
Foliage of extinct seed plants called medullosans. Characterized
by pinnules with a constricted base; pinnules often with rounded
tips.

Cordaitanthus (cone)

2 ins

Cordaites

Cordaitales

Carboniferous–Permian: Worldwide
This extinct group includes the ancestors of the living conifers. Cordaitales were a conspicuous component of the Late Paleozoic flora, and the group included large trees as well as small shrubs.

Cordaites *Carboniferous–Permian: Worldwide*
Fragment of a leaf showing the long, strap-like form characteristic of the group. Veins are parallel to the long axis of the leaf.

Cordaitanthus *Carboniferous–Permian: Worldwide*
This name is applied to either ovulate or pollen-producing cordaitalean cones. These are typically loosely constructed organs. Compare this structure with the more compact cone of *Araucaria*.

Mesozoic and Tertiary Plants

Ginkgoales

Permian–Recent: Worldwide
This group contains a single living species, *Ginkgo biloba* (SE China), but it was an important and diverse element of Mesozoic floras. Living *Ginkgo* is a large tree.

Ginkgo (maidenhair tree) *Permian–Recent: Worldwide*
The characteristic fan-shaped leaves are typically bilobed in the living species and many fossils. The leaves of some extinct species have many more lobes. Leaves have simple dichotomous veins.

Coniferales (conifers)

Triassic–Recent: Worldwide
An important living group of seed plants that includes pines and redwoods. Leaves are usually long and narrow, and seeds are borne in cones. Conifers diversified during the Triassic, and they were a major component of Jurassic and Cretaceous floras.

Araucaria *Triassic–Recent: Worldwide*
Living Araucariaceae are confined to the southern hemisphere and comprise about 40 species. *Araucaria* includes the Monkey-Puzzle Tree and the Norfolk Island Pine. Silicified fossils of A. *mirabilis* cone are shown here with helical pattern of scales, and polished section of cone with bract-bearing ovules (both shown opposite).

Sequoiadendron (giant Sequoia) *Tertiary–Recent: NA*
From California, the giant Sequoia (Taxodiaceae) is one of the largest living trees. Bears small fossil cones with relatively few scales.

Bennettitales

Triassic–Cretaceous: Worldwide
An extinct group with ovules and pollen organs grouped into elaborate, flower-like heads. The group is not closely related to living cycads despite some remarkable superficial similarities in habit and leaf morphology.

Williamsonia *Triassic–Cretaceous: Worldwide*
Flower-like cone showing radially arranged petal-like bracts.

Pterophyllum *Triassic–Recent: Worldwide*
Bennettitalean leaf resembling some cycads and ferns. The pinnules have parallel margins and veins.

Cycadales (sago palms)

2 ins

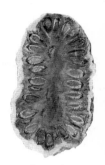

Permian–Recent: Worldwide
A small living group of seed plants comprising ten genera including the Sago Palm (*Cycas revoluta*). Male and female cones are borne on separate plants. This group is only distantly related to Bennettitales and flowering plants.

Nilssonia *Permian–Cretaceous: Worldwide*
Leaves are lanceolate or pinnate with fine parallel veins. Strongly resembling *Pterophyllum* (pictured below) and distinguishable on microscopic epidermal characters (morphology of stomates).

Araucaria

Ginkgo

Araucaria

Pterophyllum

Sequoiadendron
(cone)

Williamsonia (cone)

Nilssonia

Angiosperms (flowering plants)

Cretaceous–Recent: Worldwide
More than 80% of living land plants (ca 250,000 species) are angiosperms. They probably originated in the Triassic, but the earliest unequivocal fossils come from the Lower Cretaceous. Evidence from fossil pollen grains and leaves documents a rapid diversification of angiosperms in the Early Cretaceous, and this group is the dominant element of many floras by the early Tertiary. The most commonly collected fossils are leaves and wood. Minute, charcoalified flowers and seeds can also be found in Cretaceous clays. These represent the remains of plants that have been burnt in bush fires before fossilization. Recovery of flowers and seeds usually requires special preparation and equipment. Appropriate clay sediments are disaggregated in water, and the organic material is sieved and examined with a microscope. Some common leaves, fruits and woods are listed below.

Dicotyledons

Most flowering plants are dicotyledons, and the earliest fossil angiosperms belong to this group. Primitive living dicotyledons include Nymphaeales, Piperales, Aristolochiaceae, Winteraceae, Chloranthaceae, Calycanthaceae, Laurales and Magnoliales. Several living groups are recognizable by the Late Cretaceous (Turonian), including Lauraceae, and fossils that are probably related to Chloranthaceae, Magnoliaceae, platinoids and rosiids. Leaves of dicotyledons typically have a complex network of veins.

Laurus (laurel) *Tertiary–Recent: E Af*
Leaf with entire margins. Living Lauraceae comprise some 3,000 species of mainly tropical and subtropical trees and shrubs. This family has an extensive Tertiary fossil record and has been documented in the Cretaceous.

2 ins

Laurus

Platanus

Platanus (plane) *Tertiary–Recent: Asia E NA*
Palmately lobed leaf. Living Platanaceae contain approximately eight species of temperate and tropical trees. Leaves, wood and infructescences are first documented in the mid-Cretaceous. This family is an important component of angiosperm floras throughout the Late Cretaceous and early Tertiary.

Zelkova (caucasian elm) *Oligocene–Recent E Asia*
Leaf with one tooth per secondary vein. Living *Zelkova* (Ulmaceae) comprises six or seven species of tree. The group has an excellent European fossil record.

Rhus (varnish tree) *Tertiary–Recent NA Asia.*
Rhus is a member of the Anacardiaceae which comprise some 600 living species. Family includes the Cashew and Pistachio.

Acer (maple, sycamore) *Tertiary–Recent: Asia E NA Af*
Palmately lobed leaf with serrated margin. There are some 140 living species of Aceraceae, most of which are trees or shrubs. Characteristic winged fruits, seeds and leaves are common in the Paleocene and Oligocene. Leaves of *Acerites* are known from the Cretaceous.

Populus (poplar) *Tertiary–Recent: NA E Asia*
Ovate leaves with a crenate margin. Flowers are borne in a raceme. Poplar (and willow) is in the Salicaceae, a group containing some 536 species. Fossils resembling living Salicaceae are known from the middle Eocene.

Zelkova

2 ins

Rhus

Acer

Populus

Fossil wood

Wood is frequently preserved in the fossil record through replacement by minerals such as silicates, calcium and magnesium carbonates, and pyrite. Because of the robust nature of the plant cell wall, soft tissue preservation at the cellular level is much more common in plants than animals. Special techniques are required to cut and polish fossil wood, and observation of cell structure requires a microscope. Gymnosperm wood (conifers and their relatives) is generally homogeneous with long straight tracheid cells, and it lacks large vessel cells. Angiosperm wood is more heterogeneous, and usually contains tracheids and large vessels as well as one or more categories of fibres. Because of the attractive patterns caused by mineralization and the structure of the wood itself, polished sections of fossilized wood are frequently sold in rock shops.

Quercus (oak) *Tertiary–Recent: NA E Asia Af*
There are some 450 living species of oaks (Fagaceae), which are widely distributed in temperate and tropical regions. The polished section through this silicified trunk shows conspicuous growth rings. Growth rings in fossil wood can provide important information on paleoclimate. The presence of rings indicates growth in a seasonal climate (e.g. temperate) whereas the absence of rings suggests a non-seasonal climate (e.g. humid tropics).

Fossil fruits

Fruits are abundant in the fossil record. Soft tissue preservation at the cellular level is a common feature and can provide much information on the affinity of the fruit. Pollen attached to the stigmatic surface can sometimes be used to link fruits to particular fossil flowers.

Prosopis (mesquite) *Eocene–Recent: NA SA Af Asia*
Part of fruit (legume, Mimoseae) with six seeds visible. Living *Prosopis* comprises some 44 species of nitrogen-fixing trees that live in frost-free, arid environments.

Anonaspermum *Eocene–Recent: Worldwide*
Internal cast (pyrite) of storage tissue (endosperm) of seed is shown here. The endosperm has a characteristic and easily recognizable corrugate or punctate surface. *Anonaspermum* is a member of the Anonaceae, a family of predominantly lowland tropical trees, shrubs, and climbers.

2 ins _____   *Prosopis*

Anonaspermum

2 ins

Monocotyledons

About 22% of living angiosperms are monocotyledons, and over half of this diversity is contained in four families: Orchidaceae (orchids), Poaceae (grasses), Cyperaceae (sedges), and Arecaceae (palms). The Cretaceous fossil record of monocotyledons is poor compared to that of dicotyledons. Evidence of a rapid diversification is provided in the Late Cretaceous by fruits of Zingiberales (gingers and their allies) and the leaves and stems of palms. Many groups of monocotyledons had evolved by the early Tertiary. Leaves of monocotyledons typically have parallel veins.

Sabalites (palm) *Late Cretaceous–Recent: Worldwide*
Often large fossils, this fragment of palm leaf shows the parallel veins typical of monocotyledons. The veins are visible as ridges running along the leaf.

Nipa (palm) *Tertiary–Recent: Worldwide*
These palm fruits are often large and are common fossils in certain Eocene deposits (e.g. London Clay). Most living species of palms are plants of tropical and subtropical climates.

Palmoxylon (palm) *Late Cretaceous–Tertiary: Worldwide*
One of the most common fossil members of the palm family. This silicified stem contains evenly distributed vascular bundles. There are no growth rings in this specimen.

Palmoxylon
(cross-section of stem)

2 ins

Sabalites

Quercus (polished section
through trunk)

Nipa (fruit)

Geological Timescale

This stratigraphical column gives the time span of each geological time interval. "Tertiary" now has no formal international status but is included here as an informal "sub-era" since it is still in common usage. The Earth is actually over 4200 million years old, and this column shows only the part that can be dated readily using fossils (the Phanerozoic eon). Older geological time (the Precambrian) is divided into two major eons, the Proterozoic (2500–545 million years) and Archaean (3800–2500 million years). The Quaternary is actually divided into two epochs, the Holocene or Recent (10,000 years to present) and the Pleistocene (1.8 million years to 10,000 years). Note that a different scale is used for each era: the eras are shown to the same scale at the foot of the diagram.

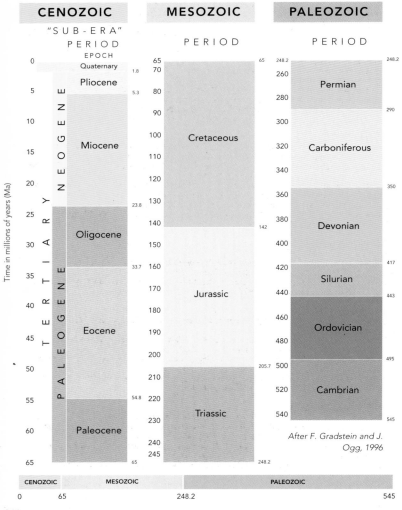

After F. Gradstein and J. Ogg, 1996

Further Reading

General geology

Duff, D. **Holmes' Principles of Physical Geology**. 4th edition. Chapman & Hall, London, 1993.

Press, F. and Siever, R. **Understanding Earth**. W. H. Freeman & Co., New York. 1994.

Skinner, B. J. and Porter, S. C. **The dynamic earth: an introduction to physical geology**. 3rd edition. John Wiley & Sons Inc., New York, 1995.

Minerals

Blackburn, W. H. and Dennen, W. H. **Encyclopedia of Mineral Names**. Edited by R. F. Martin. Canadian Mineralogist Special Publication 1, 1997.

Clark, A. M. **Hey's Mineral Index: mineral species, varieties and synonyms**. Natural History Museum Publications and Chapman & Hall, London, 1993.

Dana, E. S. A. **Textbook of Mineralogy**. 4th edition, revised and enlarged by Ford, W. E. John Wiley & Sons Inc., New York, London and Sydney, 1932.

Deer, W. A., Howie, R. A. and Zussman, J. **An Introduction to Rock-forming Minerals**. 2nd edition, Longmans, London, 1992.

Dennen, W. H. **Mineral Resources: geology, exploration and development**. Taylor and Francis, New York, 1989.

Gribble, C. D. **Rutley's Elements of Mineralogy**. 27th edition, Unwin Hyman, London, 1988.

Hall, C. **Gemstones**. The Natural History Museum, London, 1987.

Hall, C. **Gemstones**. Dorling Kindersley Ltd, London, 1994.

Klein, C. and Hurlbut, C. S. Jr. **Manual of Mineralogy** (after James D. Dana). 21st edition (revised). John Wiley & Sons Inc., New York, 1999.

Rocks

Barker, D. S. **Igneous Rocks**. Prentice-Hall, Inc., Englewood Cliffs, New Jersey, 1983.

Francis, P. **Volcanoes: a planetary perspective**. Clarendon Press, Oxford, 1993.

Hutchison, R. **The search for our beginning**. British Museum (Natural History), London and Oxford University Press, Oxford, 1983.

Hutchison, R. and Graham, A. **Meteorites**. The Natural History Museum, London, 1992.

Mason, R. **Petrology of Metamorphic Rocks**. Unwin Hyman, London, 1990.

Pettijohn, F. J. **Sedimentary Rocks**. Harper and Brothers, New York, 1949.

Van Rose, S. and Mercer, I. F. **Volcanoes**. 2nd edition. British Museum (Natural History), London, 1991.

Fossils

Guides

Murray, J. W. (editor). **Atlas of invertebrate macrofossils**. Longman (for the Palaeontological Association) Harlow, Essex, 1985.

Walker, C. and Ward, D., photography by Keates, C. **Eye Witness Handbook: Fossils**. Dorling Kindersley, London, 1992.

Taxonomic paleontology

Black, R. M. **The elements of palaeontology**. 2nd edition. Cambridge, 1988. University Press.

Boardman R. S., Cheetham A. H. and Rowel A. J. (editors). **Fossil invertebrates**. Blackwell Scientific Publications, Palo Alto, California, 1987.

Clarkson, E. N. K. **Invertebrate palaeontology and evolution**. 3rd edition. Chapman & Hall, London, 1993.

Treatise on invertebrate paleontology. University of Kansas, Lawrence, Kansas, and Geological Society of America, Boulder, Colorado. 1952 onward.

General paleontology

Cowen, R. **History of life**. 2nd edition. Blackwell Scientific Publications, Oxford, 1995.

Fortey, R. A. **Life: an unauthorised biography, a natural history of the first four thousand million years of life on earth**. Flamingo, London, 1998.

Gardom, T. with Angela Milner. **The Natural History Museum book of dinosaurs**. Virgin Books, London, 1993.

Goldring, R. **Fossils in the field; information potential and analysis**. Longman, Harlow, 1991.

McKerrow, W. S. (editor). *The ecology of fossils; an illustrated guide*. MIT Press, Cambridge, Massachusetts, 1978.

Rudwick, M. J. S. *The meaning of fossils: episodes in the history of palaeontology*. Macdonald, London, 1972.

Tassy, Pascal. *The Message of Fossils* (translated from French by N. Hartmann), McGraw Hill, New York. (First published as *Message de fossiles*, 1991), 1993.

Swinnerton, H. H. *Fossils*. Collins (New Naturalist Library 42), London, 1960.

Wendt, H. *Before the Deluge* (translated from German by R. and C. Winston). Victor Gollancz Ltd., London. (First published as *Ehe die Sintflut kam*, 1965), 1968.

Index

Page numbers in heavier type
refer to main description